Linear Algebra Suppl

to Accompany

Calculus

with Analytic Geometry

FIFTH EDITION

HOWARD ANTON

DREXEL UNIVERSITY

JOHN WILEY & SONS, INC.
New York   Chichester   Brisbane   Toronto   Singapore

ISBN 0-471-10677-1

Printed in the United States of America

10 9 8 7 6 5 4 3 2 1

Preface

This supplement, which is written to accompany Calculus with Analytic Geometry, 5th edition, by Howard Anton, is intended to provide a brief introduction to those aspects of linear algebra that are of immediate concern to the calculus student. The presentation is "methods oriented" with relatively little emphasis on proof.

Readers interested in a more complete treatment of the subject are referred to Elementary Linear Algebra, 7th edition, John Wiley & Sons, Inc., 1994, by Howard Anton.

Contents

CHAPTER 1   SYSTEMS OF LINEAR EQUATIONS AND MATRICES   1

   1.1   Introduction to Systems of Linear Equations   1

   1.2   Gauss-Jordan Elimination   10

   1.3   Homogeneous Systems of Linear Equations   22

   1.4   Matrices and Matrix Operations   27

   1.5   Rules of Matric Arithmetic   38

   1.6   A Method for Finding A^{-1} 51

   1.7   Further Results on Systems of Equations and Invertibility   56

CHAPTER 2   DETERMINANTS   65

   2.1   Definitions   65

   2.2   Properties of Determinants; Alternate Methods of
       Evaluation   74

   2.3   Adjoint Formula for A^{-1}; Cramer's Rule   85

CHAPTER 3   APPLICATIONS TO CALCULUS   95

   3.1   Euclidean N-Dimensional Space   95

   3.2   Rotation of Axes   104

   3.3   Transformations from R^n to R^m 109

ANSWERS TO EXERCISES   123

1 Systems of Linear Equations and Matrices

1.1 INTRODUCTION TO SYSTEMS OF LINEAR EQUATIONS

In this section we introduce basic terminology and discuss a method for solving systems of linear equations.

A line in the xy-plane can be represented algebraically by an equation of the form

$$a_1 x + a_2 y = b$$

An equation of this kind is called a linear equation in the variables x and y. More generally, we define a *linear equation* in the n variables $x_1, x_2, \ldots, x_n$ to be one that can be expressed in the form

$$a_1 x_1 + a_2 x_2 + \ldots + a_n x_n = b$$

where $a_1, a_2, \ldots, a_n$ and b are real constants.

Example 1

The following are linear equations:

$$x + 3y = 7 \qquad\qquad x_1 - 2x_2 - 3x_3 + x_4 = 7$$

$$y = \frac{1}{2}x + 3z + 1 \qquad\qquad x_1 + x_2 + \cdots + x_n = 1$$

Observe that a linear equation does not involve any products or roots of variables. All variables occur only to the first power and do not appear as arguments for trigonometric, logarithmic, or exponential functions. The following are *not* linear equations:

1

$$x + 3y^2 = 7 \qquad\qquad 3x + 2y - z + xz = 4$$

$$y - \sin x = 0 \qquad\qquad \sqrt{x_1} + 2x_2 + x_3 = 1$$

A *solution* of a linear equation $a_1x_1 + a_2x_2 + \cdots + a_nx_n = b$ is a sequence of n numbers $s_1, s_2, \ldots, s_n$ such that the equation is satisfied when we substitute $x_1 = s_1, x_2 = s_2, \ldots, x_n = s_n$. The set of all solutions of the equation is called its *solution set*.

Example 2

Find the solution set of each of the following:

(i) $4x - 2y = 1$ (ii) $x_1 - 4x_2 + 7x_3 = 5$

To find solutions of (i), we can assign an arbitrary value to x and solve for y, or choose an arbitrary value for y and solve for x. If we follow the first approach and assign x an arbitrary value t, we obtain

$$x = t, \qquad y = 2t - \frac{1}{2}$$

These formulas describe the solution set in terms of the arbitrary parameter t. Particular numerical solutions can be obtained by substituting specific values for t. For example, t = 3 yields the solution x = 3, y = 11/2 and t = -1/2 yields the solution x = -1/2, y = -3/2.

If we follow the second approach and assign y the arbitrary value t, we obtain

$$x = \frac{1}{2}t + \frac{1}{4}, \qquad y = t$$

Although these formulas are different from those obtained above, they yield the same solution set as t varies over all possible real numbers. For example, the previous formulas gave the solution x = 3, y = 11/2 when t = 3, while these formulas yield this solution when t = 11/2.

To find the solution set of (ii) we can assign arbitrary values to any two variables and solve for the third variable. In particular, if we assign arbitrary values s and t to x_2 and x_3, respectively, and solve for x_1, we obtain

$$x_1 = 5 + 4s - 7t, \qquad x_2 = s, \qquad x_3 = t$$

A finite set of linear equations in the variables $x_1, x_2, \ldots, x_n$ is called a *system of linear equations* or a *linear system*. A sequence of numbers $s_1, s_2, \ldots, s_n$ is called a *solution* of the system if $x_1 = s_1, x_2 = s_2, \ldots, x_n = s_n$ is a solution of every equation in the system. For example, the system

$$4x_1 - x_2 + 3x_3 = -1$$

$$3x_1 + x_2 + 9x_3 = -4$$

has the solution $x_1 = 1$, $x_2 = 2$, $x_3 = -1$ since these values satisfy both equations. However, $x_1 = 1$, $x_2 = 8$, $x_3 = 1$ is not a solution since these values satisfy only the first of the two equations in the system.

Not all systems of linear equations have solutions. For example, if we multiply the second equation of the system

$$x + y = 4$$

$$2x + 2y = 6$$

by 1/2, it becomes evident that there are no solutions, since the two equations in the resulting system

$$x + y = 4$$

$$x + y = 3$$

contradict each other.

A system of equations that has no solutions is said to be *inconsistent*. If there is at least one solution, it is called *consistent*. To illustrate the possibilities that can occur in solving systems of linear equations, consider a general system of two linear equations in the unknowns x and y:

$$a_1 x + b_1 y = c_1 \qquad (a_1, b_1 \text{ both not zero})$$

$$a_2 x + b_2 y = c_2 \qquad (a_2, b_2 \text{ both not zero})$$

The graphs of these equations are lines; call them l_1 and l_2. Since a point (x,y) lies on a line if and only if the numbers x and y satisfy the equation of the line, the solutions of the system of equations will correspond to points of intersection of l_1 and l_2. There are three possibilities (Figure 1.1).

(a) The lines l_1 and l_2 may be parallel, in which case there is no intersection, and consequently no solution to the system.

(b) The lines l_1 and l_2 may intersect at only one point, in which case the system has exactly one solution.

(c) The lines l_1 and l_2 may coincide, in which case there are infinitely many points of intersection, and consequently infinitely many solutions to the system.

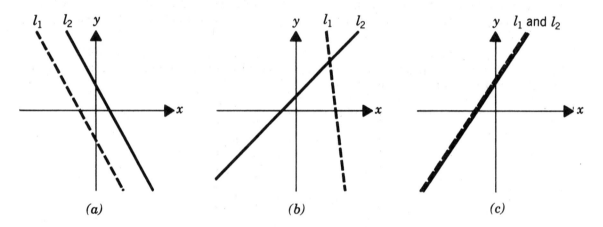

Figure 1.1   (a) No solution.   (b) One solution.   (c) Infinitely many solutions.

Although we have considered only two equations with two unknowns here, we will show later that this same result holds for arbitrary systems; that is, *every system of linear equations has either no solutions, exactly one solution, or infinitely many solutions*

An arbitrary system of m linear equations in n unknowns will be written

$$a_{11}x_1 + a_{12}x_2 + \cdots + a_{1n}x_n = b_1$$

$$a_{21}x_1 + a_{22}x_2 + \cdots + a_{2n}x_n = b_2$$

$$\vdots \qquad \vdots \qquad \qquad \vdots \qquad \vdots$$

$$a_{m1}x_1 + a_{m2}x_2 + \cdots + a_{mn}x_n = b_m$$

where $x_1, x_2, \ldots, x_n$ are the unknowns and the subscripted a's and b's denote constants.

For example, a general system of three linear equations in four unknowns will be written

$$a_{11}x_1 + a_{12}x_2 + a_{13}x_3 + a_{14}x_4 = b_1$$

$$a_{21}x_1 + a_{22}x_2 + a_{23}x_3 + a_{24}x_4 = b_2$$

$$a_{31}x_1 + a_{32}x_2 + a_{33}x_3 + a_{34}x_4 = b_3$$

The double subscripting on the coefficients of the unknowns is a useful device that we shall employ to establish the location of the coefficient in the system. The first subscript on the coefficient a_{ij} indicates the equation in which the coefficient occurs, and the second subscript indicates which unknown it multiplies. Thus, a_{12} is in the first equation and multiples unknown x_2.

If we mentally keep track of the location of the +'s, the x's, and the ='s, a system of m linear equations in n unknowns can be abbreviated by writing only the rectangular array of numbers:

$$\begin{bmatrix} a_{11} & a_{12} & \cdots & a_{1n} & b_1 \\ a_{21} & a_{22} & \cdots & a_{2n} & b_2 \\ \vdots & \vdots & & \vdots & \vdots \\ a_{m1} & a_{m2} & \cdots & a_{mn} & b_m \end{bmatrix}$$

This is called the *augmented matrix* for the system. (The term *matrix* is used in mathematics to denote a rectangular array of numbers. Matrices arise in many contexts; we shall study them in more detail in later sections.) To illustrate, the augmented matrix for the system of equations

$$x_1 + x_2 + 2x_3 = 9$$

$$2x_1 + 4x_2 - 3x_3 = 1$$

$$3x_1 + 6x_2 - 5x_3 = 0$$

is

$$\begin{bmatrix} 1 & 1 & 2 & 9 \\ 2 & 4 & -3 & 1 \\ 3 & 6 & -5 & 0 \end{bmatrix}$$

REMARK. When constructing an augmented matrix, the unknowns must be written in the same order in each equation.

The basic method for solving a system of linear equations is to replace the given system by a new system that has the same solution set, but is easier to solve. This new system is generally obtained in a series of steps by applying the following three types of operations systematically to eliminate unknowns.

1. Multiply an equation through by a nonzero constant.
2. Interchange two equations.
3. Add a multiple of one equation to another.

Since the rows (horizontal lines) in an augmented matrix correspond to the equations in the associated system, these three operations correspond to the following operations on the rows of the augmented matrix.

1. Multiple a row through by a nonzero constant.
2. Interchange two rows.
3. Add a multiple of one row to another row.

These are called *elementary row operations*. The following example illustrates how these operations can be used to solve systems of linear equations. Since a systematic procedure for finding solutions will be derived in the next section, it is not necessary to worry about how the steps in this example were selected. The main effort at this time should be devoted to understanding the computations and the discussion.

Example 3

In the left column below we solve a system of linear equations by operating on the equations in the system, and in the right column we solve the same system by operating on the rows of the augmented matrix.

$$
\begin{aligned}
x + y + 2z &= 9 \\
2x + 4y - 3z &= 1 \\
3x + 6y - 5z &= 0
\end{aligned}
\qquad
\begin{bmatrix}
1 & 1 & 2 & 9 \\
2 & 4 & -3 & 1 \\
3 & 6 & -5 & 0
\end{bmatrix}
$$

Add -2 times the first equation to the second to obtain

Add -2 times the first row to the second to obtain

$$x + y + 2z = 9$$
$$2y - 7z = -17$$
$$3x + 6y - 5z = 0$$

$$\begin{bmatrix} 1 & 1 & 2 & 9 \\ 0 & 2 & -7 & -17 \\ 3 & 6 & -5 & 0 \end{bmatrix}$$

Add -3 times the first equation to the third to obtain

Add -3 times the first row to the third to obtain

$$x + y + 2z = 9$$
$$2y - 7z = -17$$
$$3y - 11z = -27$$

$$\begin{bmatrix} 1 & 1 & 2 & 9 \\ 0 & 2 & -7 & -17 \\ 0 & 3 & -11 & -27 \end{bmatrix}$$

Multiply the second equation by 1/2 to obtain

Multiply the second row by 1/2 to obtain

$$x + y + 2z = 9$$
$$y - \frac{7}{2} z = - \frac{17}{2}$$
$$3y - 11z = -27$$

$$\begin{bmatrix} 1 & 1 & 2 & 9 \\ 0 & 1 & -\frac{7}{2} & -\frac{17}{2} \\ 0 & 3 & -11 & -27 \end{bmatrix}$$

Add -3 times the second equation to the third to obtain

Add -3 times the second row to the third to obtain

$$x + y + 2z = 9$$
$$y - \frac{7}{2} z = - \frac{17}{2}$$
$$- \frac{1}{2} z = - \frac{3}{2}$$

$$\begin{bmatrix} 1 & 1 & 2 & 9 \\ 0 & 1 & -\frac{7}{2} & -\frac{17}{2} \\ 0 & 0 & -\frac{1}{2} & -\frac{3}{2} \end{bmatrix}$$

Multiply the third equation by -2 to obtain

Multiply the third row by -2 to obtain

$$x + y + 2z = 9$$
$$y - \frac{7}{2} z = - \frac{17}{2}$$
$$z = 3$$

$$\begin{bmatrix} 1 & 1 & 2 & 9 \\ 0 & 1 & -\frac{7}{2} & -\frac{17}{2} \\ 0 & 0 & 1 & 3 \end{bmatrix}$$

Add -1 times the second equation to the first to obtain

$$x \quad + \frac{11}{2} z = \frac{35}{2}$$

$$y - \frac{7}{2} z = -\frac{17}{2}$$

$$z = 3$$

Add $-\frac{11}{2}$ times the third equation to the first and $\frac{7}{2}$ times the third equation to the second to obtain

$$x \quad = 1$$

$$y \quad = 2$$

$$z = 3$$

Add -1 times the second row to the first to obtain

$$\begin{bmatrix} 1 & 0 & \frac{11}{2} & \frac{35}{2} \\ 0 & 1 & -\frac{7}{2} & -\frac{17}{2} \\ 0 & 0 & 1 & 3 \end{bmatrix}$$

Add $-\frac{11}{2}$ times the third row to the first and $\frac{7}{2}$ times the third row to the second to obtain

$$\begin{bmatrix} 1 & 0 & 0 & 1 \\ 0 & 1 & 0 & 2 \\ 0 & 0 & 1 & 3 \end{bmatrix}$$

The solution

$$x = 1, \qquad y = 2, \qquad z = 3$$

is now evident.

EXERCISE SET 1.1

1. Which of the following are linear equations in x_1, x_2, and x_3?

(a) $x_1 + 2x_1 x_2 + x_3 = 2$

(b) $x_1 + x_2 + x_3 = \sin k$ (k is a constant)

(c) $x_1 - 3x_2 + 2x_3^{1/2} = 4$

(d) $x_1 = \sqrt{2}x_3 - x_2 + 7$

(e) $x_1 + x_2^{-1} - 3x_3 = 5$

(f) $x_1 = x_3$

2. Find the solution set of:

(a) $6x - 7y = 3$

(b) $2x_1 + 4x_2 - 7x_3 = 8$

(c) $-3x_1 + 4x_2 - 7x_3 + 8x_4 = 5$ (d) $2v - w + 3x + y - 4z = 0.$

3. Find the augmented matrix for each of the following systems of linear equations.

(a) $x_1 - 2x_2 = 0$

    $3x_1 + 4x_2 = -1$

    $2x_1 - x_2 = 3$

(b) $x_1 + x_3 = 1$

    $-x_1 + 2x_2 - x_3 = 3$

(c) $x_1 + x_3 = 1$

    $2x_2 - x_3 + x_5 = 2$

    $2x_3 + x_4 = 3$

(d) $x_1 = 1$

    $x_2 = 2$

4. Find a system of linear equations corresponding to each of the following augmented matrices.

(a) $\begin{bmatrix} 1 & 0 & -1 & 2 \\ 2 & 1 & 1 & 3 \\ 0 & -1 & 2 & 4 \end{bmatrix}$

(b) $\begin{bmatrix} 1 & 0 & 0 \\ 0 & 1 & 0 \\ 1 & -1 & 1 \end{bmatrix}$

(c) $\begin{bmatrix} 1 & 2 & 3 & 4 & 5 \\ 5 & 4 & 3 & 2 & 1 \end{bmatrix}$

(d) $\begin{bmatrix} 1 & 0 & 0 & 0 & 1 \\ 0 & 1 & 0 & 0 & 2 \\ 0 & 0 & 1 & 0 & 3 \\ 0 & 0 & 0 & 1 & 4 \end{bmatrix}$

5. For which value(s) of the constant k does the following system of linear equations have no solutions? Exactly one solution? Infinitely many solutions?

$$x - y = 3$$
$$2x - 2y = k$$

6. Consider the system of equations

$$ax + by = k$$

$$cx + dy = \ell$$

$$ex + fy = m$$

Discuss the relative positions of the lines $ax + by = k$, $cx + dy = \ell$, and $ex + fy = m$ when:

(a) the system has no solutions

(b) the system has exactly one solution

(c) the system has infinitely many solutions.

7. Show that if the system of equations in Exercise 6 is consistent, then at least one equation can be discarded from the system without altering the solution set.

8. Let $k = \ell = m = 0$ in Exercise 6; show that the system must be consistent. What can be said about the point of intersection of the three lines if the system has exactly one solution?

9. Consider the system of equations

$$x + y + 2z = a$$

$$x \quad + \ z = b$$

$$2x + y + 3z = c$$

Show that in order for this system to be consistent, a, b, and c must satisfy $c = a + b$.

10. Prove: If the linear equations $x_1 + kx_2 = c$ and $x_1 + \ell x_2 = d$ have the same solution set, then the equations are identical.

1.2 GAUSS-JORDAN ELIMINATION

In this section we give a systematic procedure for solving systems of linear equations; it is based on the idea of reducing the augmented matrix to a form that is simple enough so that the system of equations can be solved by inspection.

In the last step of Example 3 in the previous section we obtained the augmented matrix

$$\begin{bmatrix} 1 & 0 & 0 & 1 \\ 0 & 1 & 0 & 2 \\ 0 & 0 & 1 & 3 \end{bmatrix} \qquad (1)$$

from which the solution of the system was evident.

Matrix (1) is an example of a matrix that is in *reduced row-echelon form*. To be of this form, a matrix must have the following properties.

1. If a row does not consist entirely of zeros, then the first nonzero number in the row is a 1. (We call this a *leading 1*.)
2. If there are any rows that consist entirely of zeros, then they are grouped together at the bottom of the matrix.
3. In any two successive rows that do not consist entirely of zeros, the leading 1 in the lower row occurs farther to the right than the leading 1 in the higher row.
4. Each column that contains a leading 1 has zeros everywhere else.

A matrix having properties 1, 2, and 3 is said to be in *row-echelon form*.

Example 1

The following matrices are in reduced row-echelon form.

$$\begin{bmatrix} 1 & 0 & 0 & 4 \\ 0 & 1 & 0 & 7 \\ 0 & 0 & 1 & -1 \end{bmatrix} \begin{bmatrix} 1 & 0 & 0 \\ 0 & 1 & 0 \\ 0 & 0 & 1 \end{bmatrix} \begin{bmatrix} 0 & 1 & -2 & 0 & 1 \\ 0 & 0 & 0 & 1 & 3 \\ 0 & 0 & 0 & 0 & 0 \\ 0 & 0 & 0 & 0 & 0 \end{bmatrix} \begin{bmatrix} 0 & 0 \\ 0 & 0 \end{bmatrix}$$

The following matrices are in row-echelon form.

$$\begin{bmatrix} 1 & 4 & 3 & 7 \\ 0 & 1 & 6 & 2 \\ 0 & 0 & 1 & 5 \end{bmatrix} \begin{bmatrix} 1 & 1 & 0 \\ 0 & 1 & 0 \\ 0 & 0 & 0 \end{bmatrix} \begin{bmatrix} 0 & 1 & 2 & 6 & 0 \\ 0 & 0 & 1 & -1 & 0 \\ 0 & 0 & 0 & 0 & 1 \end{bmatrix}$$

The reader should check to see that each of the above matrices satisfied all the necessary requirements.

REMARK. A matrix in row-echelon form must have zeros below each leading 1, and a matrix in reduced row-echelon form must have zeros above and below each leading 1.

If, by a sequence of elementary row operations, the augmented matrix for a system of linear equations is put in reduced row-echelon form, then the solution set for the system can be obtained by inspection or after a few simple steps. The next example illustrates this point.

Example 2

Suppose that the augmented matrix for a system of linear equations has been reduced by row operations to the given reduced row-echelon form. Solve the system.

(a) $\begin{bmatrix} 1 & 0 & 0 & 5 \\ 0 & 1 & 0 & -2 \\ 0 & 0 & 1 & 4 \end{bmatrix}$
(b) $\begin{bmatrix} 1 & 0 & 0 & 4 & -1 \\ 0 & 1 & 0 & 2 & 6 \\ 0 & 0 & 1 & 3 & 2 \end{bmatrix}$

(c) $\begin{bmatrix} 1 & 6 & 0 & 0 & 4 & -2 \\ 0 & 0 & 1 & 0 & 3 & 1 \\ 0 & 0 & 0 & 1 & 5 & 2 \\ 0 & 0 & 0 & 0 & 0 & 0 \end{bmatrix}$
(d) $\begin{bmatrix} 1 & 0 & 0 & 0 \\ 0 & 1 & 2 & 0 \\ 0 & 0 & 0 & 1 \end{bmatrix}$

Solution to (a). The corresponding system of equations is

$$x_1 = 5$$
$$x_2 = -2$$
$$x_3 = 4$$

By inspection, $x_1 = 5$, $x_2 = -2$, $x_3 = 4$.

Solution to (b). The corresponding system of equations is

$$x_1 + 4x_4 = -1$$
$$x_2 + 2x_4 = 6$$
$$x_3 + 3x_4 = 2$$

Since x_1, x_2, and x_3 correspond to leading 1's in the augmented matrix, we call them *leading variables*. Solving for the leading variables in terms of x_4 gives

$$x_1 = -1 - 4x_4$$

$$x_2 = 6 - 2x_4$$

$$x_3 = 2 - 3x_4$$

Since x_4 can be assigned an arbitrary value, say t, we have infinitely many solutions. The solution set is given by the formulas

$$x_1 = -1 - 4t, \qquad x_2 = 6 - 2t, \qquad x_3 = 2 - 3t, \qquad x_4 = t$$

Solution to (c). The corresponding system of equations is

$$x_1 + 6x_2 \qquad + 4x_5 = -2$$

$$x_3 \qquad + 3x_5 = 1$$

$$x_4 + 5x_5 = 2$$

Here the leading variables are x_1, x_3, and x_4. Solving for the leading variables in terms of the remaining variable gives

$$x_1 = -2 - 4x_5 - 6x_2$$

$$x_3 = 1 - 3x_5$$

$$x_4 = 2 - 5x_5$$

Since x_5 can be assigned an arbitrary value, t, and x_2 can be assigned an arbitrary value, s, there are infinitely many solutions. The solution set is given by the formulas

$$x_1 = -2 - 4t - 6s, \qquad x_2 = s, \qquad x_3 = 1 - 3t, \qquad x_4 = 2 - 5t, \qquad x_5 = t$$

Solution to (d). The last equation in the corresponding system of equations is

$$0x_1 + 0x_2 + 0x_3 = 1$$

Since this equation can never be satisfied, there is no solution to the system.

We have just seen how easy it is to solve a system of linear equations, once its augmented matrix is in reduced row-echelon form. Now we shall give a step-by-step procedure, called *Gauss-Jordan elimination,** which can be used to reduce any matrix to this form. We shall illustrate the steps in the procedure by reducing the following matrix to reduced row-echelon form.

$$\begin{bmatrix} 0 & 0 & -2 & 0 & 7 & 12 \\ 2 & 4 & -10 & 6 & 12 & 28 \\ 2 & 4 & -5 & 6 & -5 & -1 \end{bmatrix}$$

Step 1. Locate the leftmost column (vertical line) that does not consist entirely of zeros.

$$\begin{bmatrix} 0 & 0 & -2 & 0 & 7 & 12 \\ 2 & 4 & -10 & 6 & 12 & 28 \\ 2 & 4 & -5 & 6 & -5 & -1 \end{bmatrix}$$

↑
Leftmost nonzero column

Step 2. Interchange the top row with another row, if necessary, to bring a nonzero entry to the top of the column found in Step 1.

$$\begin{bmatrix} 2 & 4 & -10 & 6 & 12 & 28 \\ 0 & 0 & -2 & 0 & 7 & 12 \\ 2 & 4 & -5 & 6 & -5 & -1 \end{bmatrix}$$

> The first and second rows in the previous matrix were interchanged.

Carl Friedrich Gauss (1777-1855). Sometimes called the "prince of mathematicians," Gauss made profound contributions to number theory, theory of functions, probability, and statistics. He discovered a way to calculate the orbits of asteroids, made basic discoveries in electromagnetic theory, and invented a telegraph.

Camille Jordan (1838-1922). Jordan was a professor at the École Polytechnique in Paris. He did pioneering work in several branches of mathematics, including matrix theory. He is particularly famous for the Jordan Curve Theorem, which states: A simple closed curve (such as a circle or a square) divides the plan into two nonintersecting connected regions.

Step 3. If the entry that is now at the top of the column found in Step 1 is a, multiply the first row by 1/a in order to introduce a leading 1.

$$\begin{bmatrix} 1 & 2 & -5 & 3 & 6 & 14 \\ 0 & 0 & -2 & 0 & 7 & 12 \\ 2 & 4 & -5 & 6 & -5 & -1 \end{bmatrix}$$

The first row of the previous matrix was multiplied by 1/2.

Step 4. Add suitable multiples of the top row to the rows below so that all entries below the leading 1 become zeros.

$$\begin{bmatrix} 1 & 2 & -5 & 3 & 6 & 14 \\ 0 & 0 & -2 & 0 & 7 & 12 \\ 0 & 0 & 5 & 0 & -17 & -29 \end{bmatrix}$$

-2 times the first row of the previous matrix was added to the third row.

Step 5. Now cover the top row in the matrix and begin again with Step 1 applied to the submatrix that remains. Continue in this way until the *entire* matrix is in row-echelon form.

$$\begin{bmatrix} 1 & 2 & -5 & 3 & 6 & 14 \\ 0 & 0 & -2 & 0 & 7 & 12 \\ 0 & 0 & 5 & 0 & -17 & -29 \end{bmatrix}$$

└─ Leftmost nonzero column in the submatrix

$$\begin{bmatrix} 1 & 2 & -5 & 3 & 6 & 14 \\ 0 & 0 & 1 & 0 & -\frac{7}{2} & -6 \\ 0 & 0 & 5 & 0 & -17 & -29 \end{bmatrix}$$

The first row in the submatrix was multiplied by -1/2 to introduce a leading 1.

$$\begin{bmatrix} 1 & 2 & -5 & 3 & 6 & 14 \\ 0 & 0 & 1 & 0 & -\frac{7}{2} & -6 \\ 0 & 0 & 0 & 0 & \frac{1}{2} & 1 \end{bmatrix}$$

-5 times the first row of the submatrix was added to the second row of the submatrix to introduce a zero below the leading 1.

$$\begin{bmatrix} 1 & 2 & -5 & 3 & 6 & 14 \\ 0 & 0 & 1 & 0 & -\frac{7}{2} & -6 \\ 0 & 0 & 0 & 0 & \frac{1}{2} & 1 \end{bmatrix}$$

The top row in the submatrix was covered and we returned again to Step 1.

↑ Leftmost nonzero column in the new submatrix

$$\begin{bmatrix} 1 & 2 & -5 & 3 & 6 & 14 \\ 0 & 0 & 1 & 0 & -\frac{7}{2} & -6 \\ 0 & 0 & 0 & 0 & 1 & 2 \end{bmatrix}$$

The first (and only) row in the new submatrix was multiplied by 2 to introduce a leading 1.

The *entire* matrix is now in row-echelon form. To find the reduced row-echelon form we need the following additional step.

Step 6. Beginning with the last nonzero row and working upward, add suitable multiples of each row to the rows above to introduce zeros above the leading 1's.

$$\begin{bmatrix} 1 & 2 & -5 & 3 & 6 & 14 \\ 0 & 0 & 1 & 0 & 0 & 1 \\ 0 & 0 & 0 & 0 & 1 & 2 \end{bmatrix}$$

7/2 times the third row of the previous matrix was added to the second row.

$$\begin{bmatrix} 1 & 2 & -5 & 3 & 0 & 2 \\ 0 & 0 & 1 & 0 & 0 & 1 \\ 0 & 0 & 0 & 0 & 1 & 2 \end{bmatrix}$$

-6 times the third row was added to the first row.

$$\begin{bmatrix} 1 & 2 & 0 & 3 & 0 & 7 \\ 0 & 0 & 1 & 0 & 0 & 1 \\ 0 & 0 & 0 & 0 & 1 & 2 \end{bmatrix}$$

5 times the second row was added to the first row.

The last matrix is in reduced row-echelon form.

Example 3

Solve by Gauss-Jordan elimination.

$$
\begin{aligned}
x_1 + 3x_2 - 2x_3 + 2x_5 &= 0 \\
2x_1 + 6x_2 - 5x_3 - 2x_4 + 4x_5 - 3x_6 &= -1 \\
5x_3 + 10x_4 + 15x_6 &= 5 \\
2x_1 + 6x_2 + 8x_4 + 4x_5 + 18x_6 &= 6
\end{aligned}
$$

The augmented matrix for the system is

$$
\begin{bmatrix}
1 & 3 & -2 & 0 & 2 & 0 & 0 \\
2 & 6 & -5 & -2 & 4 & -3 & -1 \\
0 & 0 & 5 & 10 & 0 & 15 & 5 \\
2 & 6 & 0 & 8 & 4 & 18 & 6
\end{bmatrix}
$$

Adding -2 times the first row to the second and fourth rows gives

$$
\begin{bmatrix}
1 & 3 & -2 & 0 & 2 & 0 & 0 \\
0 & 0 & -1 & -2 & 0 & -3 & -1 \\
0 & 0 & 5 & 10 & 0 & 15 & 5 \\
0 & 0 & 4 & 8 & 0 & 18 & 6
\end{bmatrix}
$$

Multiplying the second row by -1 and then adding -5 times the second row to the third row and -4 times the second row to the fourth row gives

$$
\begin{bmatrix}
1 & 3 & -2 & 0 & 2 & 0 & 0 \\
0 & 0 & 1 & 2 & 0 & 3 & 1 \\
0 & 0 & 0 & 0 & 0 & 0 & 0 \\
0 & 0 & 0 & 0 & 0 & 6 & 2
\end{bmatrix}
$$

Interchanging the third and fourth rows and then multiplying the third row of the resulting matrix by 1/6 gives the row-echelon form

$$
\begin{bmatrix}
1 & 3 & -2 & 0 & 2 & 0 & 0 \\
0 & 0 & 1 & 2 & 0 & 3 & 1 \\
0 & 0 & 0 & 0 & 0 & 1 & \dfrac{1}{3} \\
0 & 0 & 0 & 0 & 0 & 0 & 0
\end{bmatrix}
$$

Adding -3 times the third row to the second row and then adding 2 times the second row of the resulting matrix to the first row yields the reduced row-echelon form

$$\begin{bmatrix} 1 & 3 & 0 & 4 & 2 & 0 & 0 \\ 0 & 0 & 1 & 2 & 0 & 0 & 0 \\ 0 & 0 & 0 & 0 & 0 & 1 & \frac{1}{3} \\ 0 & 0 & 0 & 0 & 0 & 0 & 0 \end{bmatrix}$$

The corresponding system of equations is

$$x_1 + 3x_2 \quad + 4x_4 + 2x_5 \quad = 0$$
$$x_3 + 2x_4 \quad = 0$$
$$x_6 = \frac{1}{3}$$

(We have discarded the last equation, $0x_1 + 0x_2 + 0x_3 + 0x_4 + 0x_5 + 0x_6 = 0$, since it will be satisfied automatically by the solutions of the remaining equations.) Solving for the leading variables, we obtain

$$x_1 = -3x_2 - 4x_4 - 2x_5$$
$$x_3 = -2x_4$$
$$x_6 = \frac{1}{3}$$

If we assign x_2, x_4, and x_5 the arbitrary values r, s, and t, respectively, the solution set is given by the formulas

$$x_1 = -3r - 4s - 2t, \quad x_2 = r, \quad x_3 = -2s, \quad x_4 = s, \quad x_5 = t, \quad x_6 = \frac{1}{3}$$

REMARK. The procedure we have given for reducing a matrix to reduced row-echelon form is well suited for computer computation because it is systematic. However, this procedure sometimes introduces fractions, which might otherwise be avoided by varying the steps in the right way. Thus once the basic procedure has been mastered, the reader may wish to vary the steps in specific problems to avoid fractions (see Exercise 17). It can be proved, although we shall not do it, that no matter how the elementary row operations are varied, one will always arrive at the same reduced row-echelon form; that is, the reduced row-echelon form is unique. However, a row-echelon form is *not* unique; by changing the sequence of elementary row operations it is possible to arrive at a different row-echelon form (see Exercise 18).

EXERCISE SET 1.2

1. Which of the following are in reduced row-echelon form?

(a) $\begin{bmatrix} 1 & 0 & 0 \\ 0 & 0 & 0 \\ 0 & 0 & 1 \end{bmatrix}$
(b) $\begin{bmatrix} 0 & 1 & 0 \\ 1 & 0 & 0 \\ 0 & 0 & 0 \end{bmatrix}$
(c) $\begin{bmatrix} 1 & 1 & 0 \\ 0 & 1 & 0 \\ 0 & 0 & 0 \end{bmatrix}$

(d) $\begin{bmatrix} 1 & 2 & 0 & 3 & 0 \\ 0 & 0 & 1 & 1 & 0 \\ 0 & 0 & 0 & 0 & 1 \\ 0 & 0 & 0 & 0 & 0 \end{bmatrix}$
(e) $\begin{bmatrix} 1 & 0 & 0 & 5 \\ 0 & 0 & 1 & 3 \\ 0 & 1 & 0 & 4 \end{bmatrix}$
(f) $\begin{bmatrix} 1 & 0 & 3 & 1 \\ 0 & 1 & 2 & 4 \end{bmatrix}$

2. Which of the following are in row-echelon form?

(a) $\begin{bmatrix} 1 & 2 & 3 \\ 0 & 0 & 0 \\ 0 & 0 & 1 \end{bmatrix}$
(b) $\begin{bmatrix} 1 & -7 & 5 & 5 \\ 0 & 1 & 3 & 2 \end{bmatrix}$
(c) $\begin{bmatrix} 1 & 1 & 0 \\ 0 & 1 & 0 \\ 0 & 0 & 0 \end{bmatrix}$

(d) $\begin{bmatrix} 1 & 3 & 0 & 2 & 0 \\ 1 & 0 & 2 & 2 & 0 \\ 0 & 0 & 0 & 0 & 1 \\ 0 & 0 & 0 & 0 & 0 \end{bmatrix}$
(e) $\begin{bmatrix} 2 & 3 & 4 \\ 0 & 1 & 2 \\ 0 & 0 & 3 \end{bmatrix}$
(f) $\begin{bmatrix} 0 & 0 & 0 \\ 0 & 0 & 0 \\ 0 & 0 & 0 \end{bmatrix}$

In Exercises 3-6, suppose that the augmented matrix for a system of linear equations has been reduced by row operations to the given reduced row-echelon form. Solve the system.

3. $\begin{bmatrix} 1 & 0 & 0 & 4 \\ 0 & 1 & 0 & 3 \\ 0 & 0 & 1 & 2 \end{bmatrix}$

4. $\begin{bmatrix} 1 & 0 & 0 & 3 & 2 \\ 0 & 1 & 0 & -1 & 4 \\ 0 & 0 & 1 & 1 & 2 \end{bmatrix}$

5. $\begin{bmatrix} 1 & 5 & 0 & 0 & 5 & -1 \\ 0 & 0 & 1 & 0 & 3 & 1 \\ 0 & 0 & 0 & 1 & 4 & 2 \\ 0 & 0 & 0 & 0 & 0 & 0 \end{bmatrix}$

6. $\begin{bmatrix} 1 & 2 & 0 & 0 \\ 0 & 0 & 1 & 0 \\ 0 & 0 & 0 & 1 \end{bmatrix}$

In Exercises 7-14, solve the system by Gauss-Jordan elimination.

7. $\begin{aligned} x_1 + x_2 + 2x_3 &= 8 \\ -x_1 - 2x_2 + 3x_3 &= 1 \\ 3x_1 - 7x_2 + 4x_3 &= 10 \end{aligned}$

8. $\begin{aligned} 2x_1 + 2x_2 + 2x_3 &= 0 \\ -2x_1 + 5x_2 + 2x_3 &= 0 \\ -7x_1 + 7x_2 + x_3 &= 0 \end{aligned}$

9. $\begin{aligned} x - y + 2z - w &= -1 \\ 2x + y - 2z - 2w &= -2 \\ -x + 2y - 4z + w &= 1 \\ 3x \qquad\quad - 3w &= -3 \end{aligned}$

10. $\begin{aligned} 2x_1 - 3x_2 &= -2 \\ 2x_1 + x_2 &= 1 \\ 3x_1 + 2x_2 &= 1 \end{aligned}$

11. $\begin{aligned} 3x_1 + 2x_2 - x_3 &= -15 \\ 5x_1 + 3x_2 + 2x_3 &= 0 \\ 3x_1 + x_2 + 3x_3 &= 11 \\ 11x_1 + 7x_2 &= -30 \end{aligned}$

12. $\begin{aligned} 4x_1 - 8x_2 &= 12 \\ 3x_1 - 6x_2 &= 9 \\ -2x_1 + 4x_2 &= -6 \end{aligned}$

13. $\begin{aligned} 5x_1 + 2x_2 + 6x_3 &= 0 \\ -2x_1 + x_2 + 3x_3 &= 0 \end{aligned}$

14. $\begin{aligned} x_1 - 2x_2 + x_3 - 4x_4 &= 1 \\ x_1 + 3x_2 + 7x_3 + 2x_4 &= 2 \\ x_1 - 12x_2 - 11x_3 - 16x_4 &= 5 \end{aligned}$

15. Solve the following systems, where a, b, and c are constants.

(a) $\begin{aligned} 2x + y &= a \\ 3x + 6y &= b \end{aligned}$

(b) $\begin{aligned} x_1 + x_2 + x_3 &= a \\ 2x_1 \qquad + 2x_3 &= b \\ 3x_2 + 3x_3 &= c \end{aligned}$

16. For which values of a will the following system have no solutions?
     Exactly one solution? Infinitely many solutions?

$$x + 2y - \quad\quad 3z = \quad 4$$
$$3x - y + \quad\quad 5z = \quad 2$$
$$4x + y + (a^2 - 14)z = a + 2$$

17. Reduce

$$\begin{bmatrix} 2 & 1 & 3 \\ 0 & -2 & 7 \\ 3 & 4 & 5 \end{bmatrix}$$

to reduced row-echelon form without introducing any fractions.

18. Find two different row-echelon forms of

$$\begin{bmatrix} 1 & 3 \\ 2 & 7 \end{bmatrix}$$

19. Solve the following system of nonlinear equations for the unknown angles
     α, β, and γ, where $0 \leq \alpha \leq 2\pi$, $0 \leq \beta \leq 2\pi$, and $0 \leq \gamma < \pi$.

$$2 \sin \alpha - \quad \cos \beta + 3 \tan \gamma = 3$$
$$4 \sin \alpha + 2 \cos \beta - 2 \tan \gamma = 2$$
$$6 \sin \alpha - 3 \cos \beta + \quad \tan \gamma = 9$$

20. Describe the possible reduced row-echelon forms of

$$\begin{bmatrix} a & b & c \\ d & e & f \\ g & h & i \end{bmatrix}$$

21. Show that if $ad - bc \neq 0$, then the reduced row-echelon form of

$$\begin{bmatrix} a & b \\ c & d \end{bmatrix} \quad \text{is} \quad \begin{bmatrix} 1 & 0 \\ 0 & 1 \end{bmatrix}$$

22. Use Exercise 21 to show that if $ad - bc \neq 0$, then the system

$$ax + by = k$$
$$cx + dy = \ell$$

has exactly one solution.

1.3 HOMOGENEOUS SYSTEMS OF LINEAR EQUATIONS

As we have already pointed out, every system of linear equations has either one solution, infinitely many solutions, or no solutions at all. As we progress, there will be situations in which we will not be interested in finding solutions to a given system, but instead will be concerned with deciding how many solutions the system has. In this section we consider several cases in which it is possible to make statements about the number of solutions by inspection.

A system of linear equations is said to be *homogeneous* if all the constant terms are zero; that is, the system has the form

$$a_{11}x_1 + a_{12}x_2 + \cdots + a_{1n}x_n = 0$$
$$a_{21}x_1 + a_{22}x_2 + \cdots + a_{2n}x_n = 0$$
$$\vdots \qquad \vdots \qquad \qquad \vdots \qquad \vdots$$
$$a_{m1}x_1 + a_{m2}x_2 + \cdots + a_{mn}x_n = 0$$

Every homogeneous system of linear equations is consistent, since $x_1 = 0$, $x_2 = 0, \ldots, x_n = 0$ is always a solution. This solution is called the *trivial solution*; if there are other solutions, they are called *nontrivial solutions*.

Since a homogeneous system of linear equations must be consistent, there is either one solution or infinitely many solutions. Since one of these solutions is the trivial solution, we can make the following statement.

For a homogeneous system of linear equations, exactly one of the following is true.

1. *The system has only the trivial solution.*
2. *The system has infinitely many nontrivial solutions in addition to the trivial solution.*

There is one case in which a homogeneous system is assured of having nontrivial solutions; namely, whenever the system involves more unknowns than equations. To see why, consider the following example of four equations in five unknowns.

Example 1

Solve the following homogeneous system of linear equations by Gauss-Jordan elimination.

$$
\begin{aligned}
2x_1 + 2x_2 - x_3 + x_5 &= 0 \\
-x_1 - x_2 + 2x_3 - 3x_4 + x_5 &= 0 \\
x_1 + x_2 - 2x_3 - x_5 &= 0 \\
x_3 + x_4 + x_5 &= 0
\end{aligned}
\tag{1}
$$

The augmented matrix for the system is

$$
\begin{bmatrix}
2 & 2 & -1 & 0 & 1 & 0 \\
-1 & -1 & 2 & -3 & 1 & 0 \\
1 & 1 & -2 & 0 & -1 & 0 \\
0 & 0 & 1 & 1 & 1 & 0
\end{bmatrix}
$$

Reducing this matrix to reduced row-echelon form, we obtain

$$
\begin{bmatrix}
1 & 1 & 0 & 0 & 1 & 0 \\
0 & 0 & 1 & 0 & 1 & 0 \\
0 & 0 & 0 & 1 & 0 & 0 \\
0 & 0 & 0 & 0 & 0 & 0
\end{bmatrix}
$$

The corresponding system of equations is

$$
\begin{aligned}
x_1 + x_2 + x_5 &= 0 \\
x_3 + x_5 &= 0 \\
x_4 &= 0
\end{aligned}
\tag{2}
$$

Solving for the leading variables yields

$$
\begin{aligned}
x_1 &= -x_2 - x_5 \\
x_3 &= -x_5 \\
x_4 &= 0
\end{aligned}
$$

The solution set is therefore given by

$$x_1 = -s - t, \qquad x_2 = s, \qquad x_3 = -t, \qquad x_4 = 0, \qquad x_5 = t$$

Note that the trivial solution is obtained when $s = t = 0$.

Example 1 illustrates two important points about solving homogeneous systems of linear equations. First, none of the three elementary row operations can alter the final column of zeros in the augmented matrix, so that the system of equations corresponding to the reduced row-echelon form of the augmented matrix must also be a homogeneous system (see system 2 in Example 1). Second, depending on whether the reduced row-echelon form of the augmented matrix has any zero rows, the number of equations in the reduced system is the same or less than the number of equations in the original system (compare systems 1 and 2 in Example 1). Therefore, if the given homogeneous system has m equations in n unknowns with $m < n$, and if there are r nonzero rows in the reduced row-echelon form of the augmented matrix, we will have $r < n$. Thus, the system of equations corresponding to the reduced row-echelon form of the augmented matrix will look like

$$
\begin{aligned}
\cdots x_{k_1} \qquad\qquad\qquad + \Sigma(\ ) &= 0 \\
\cdots x_{k_2} \qquad\qquad + \Sigma(\ ) &= 0 \\
\cdots \qquad\qquad\qquad \vdots \\
x_{k_r} + \Sigma(\ ) &= 0
\end{aligned}
\tag{3}
$$

where x_{k_1}, x_{k_2}, $\ldots$, x_{k_r} are the leading variables and $\Sigma(\ )$ denotes sums that involve the $n - r$ remaining variables. Solving for the leading variables gives

$$
\begin{aligned}
x_{k_1} &= -\Sigma(\ ) \\
x_{k_2} &= -\Sigma(\ ) \\
&\vdots \\
x_{k_r} &= -\Sigma(\ )
\end{aligned}
$$

As in Example 1, we can assign arbitrary values to the variables on the right-hand side and thus obtain infinitely many solutions to the system.

In summary, we have the following important theorem.

Theorem 1. *A homogeneous system of linear equations with more unknowns than equations has infinitely many solutions.*

REMARK. Note that Theorem 1 applies only to homogeneous systems. A non-homogeneous system with more unknowns than equations need not be consistent (Exercise 11); however, if the system is consistent, it will have infinitely many solutions. We omit the proof.

EXERCISE SET 1.3

1. Without using pencil and paper, determine which of the following homogeneous systems have nontrivial solutions.

(a)
$$x_1 + 3x_2 + 5x_3 + x_4 = 0$$
$$4x_1 - 7x_2 - 3x_3 - x_4 = 0$$
$$3x_1 + 2x_2 + 7x_3 + 8x_4 = 0$$

(b)
$$x_1 + 2x_2 + 3x_3 = 0$$
$$x_2 + 4x_3 = 0$$
$$5x_3 = 0$$

(c)
$$a_{11}x_1 + a_{12}x_2 + a_{13}x_3 = 0$$
$$a_{21}x_1 + a_{22}x_2 + a_{23}x_3 = 0$$

(d)
$$x_1 + x_2 = 0$$
$$2x_1 + 2x_2 = 0$$

In Exercises 2-5 solve the given homogeneous system of linear equations.

2.
$$2x_1 + x_2 + 3x_3 = 0$$
$$x_1 + 2x_2 = 0$$
$$x_2 + x_3 = 0$$

3.
$$3x_1 + x_2 + x_3 + x_4 = 0$$
$$5x_1 - x_2 + x_3 - x_4 = 0$$

4.
$$2x_1 - 4x_2 + x_3 + x_4 = 0$$
$$x_1 - 5x_2 + 2x_3 = 0$$
$$- 2x_2 - 2x_3 - x_4 = 0$$
$$x_1 + 3x_2 + x_4 = 0$$
$$x_1 - 2x_2 - x_3 + x_4 = 0$$

5.
$$x + 6y - 2z = 0$$
$$2x - 4y + z = 0$$

6. For which value(s) of λ does the following system of equations have nontrivial solutions?

$$(\lambda - 3)x + \qquad y = 0$$
$$x + (\lambda - 3)y = 0$$

7. Consider the system of equations

$$ax + by = 0$$
$$cx + dy = 0$$
$$ex + fy = 0$$

Discuss the relative positions of the lines $ax + by = 0$, $cx + dy = 0$, and $ex + fy = 0$ when:

(a) the system has only the trivial solution

(b) the system has nontrivial solutions.

8. Consider the system of equations

$$ax + by = 0$$
$$cx + dy = 0$$

(a) Show that if $x = x_0$, $y = y_0$ is any solution and k is any constant, then $x = kx_0$, $y = ky_0$ is also a solution.

(b) Show that if $x = x_0$, $y = y_0$ and $x = x_1$, $y = y_1$ are any two solutions, then $x = x_0 + x_1$, $y = y_0 + y_1$ is also a solution.

9. Consider the systems of equations

$$(\text{I}) \quad ax + by = k \qquad (\text{II}) \quad ax + by = 0$$
$$cx + dy = \ell \qquad\qquad cx + dy = 0$$

(a) Show that if $x = x_1$, $y = y_1$, and $x = x_2$, $y = y_2$ are both solutions of I, then $x = x_1 - x_2$, $y = y_1 - y_2$ is a solution of II.

(b) Show that if $x = x_1$, $y = y_1$ is a solution of I and $x = x_0$, $y = y_0$ is a solution of II, then $x = x_1 + x_0$, $y = y_1 + y_0$ is a solution of I.

10. (a) In the system of equations numbered (3), explain why it would be incorrect to denote the leading variables by x_1, x_2, ..., x_r rather than x_{k_1}, x_{k_2}, ..., x_{k_r} as we have done.

    (b) The system of equations numbered (2) is a specific case of (3). What value does r have in this case? What are x_{k_1}, x_{k_2}, ..., x_{k_r} in this case? Write out the sums denoted by $\Sigma(\ )$ in (3).

11. Find an inconsistent linear system that has more unknowns than equations.

1.4 MATRICES AND MATRIX OPERATIONS

Rectangular arrays of real numbers arise in many contexts other than as augmented matrices for systems of linear equations. In this section we consider such arrays as objects in their own right and develop some of their properties for use in our later work.

DEFINITION. A *matrix* is a rectangular array of numbers. The numbers in the array are called the *entries* in the matrix.

Example 1

The following are matrices.

$$\begin{bmatrix} 1 & 2 \\ 3 & 0 \\ -1 & 4 \end{bmatrix} \qquad [2 \quad 1 \quad 0 \quad -3] \qquad \begin{bmatrix} -\sqrt{2} & \pi & e \\ 3 & \frac{1}{2} & 0 \\ 0 & 0 & 0 \end{bmatrix} \qquad \begin{bmatrix} 1 \\ 3 \end{bmatrix} \qquad [4]$$

As these examples indicate, matrices vary in size. The *size* of a matrix is described by specifying the number of *rows* (horizontal lines) and *columns* (vertical lines) that occur in the matrix. The first matrix in Example 1 has 3 rows and 2 columns so that its size is 3 by 2 (written 3×2). The first number always indicates the number of rows and the second indicates the number of columns. The remaining matrices in Example 1 thus have sizes 1×4, 3×3, 2×1, and 1×1, respectively.

We shall use capital letters to denote matrices and lowercase letters to denote numerical quantities; thus we might write

$$A = \begin{bmatrix} 2 & 1 & 7 \\ 3 & 4 & 2 \end{bmatrix} \quad \text{or} \quad C = \begin{bmatrix} a & b & c \\ d & e & f \end{bmatrix}$$

When discussing matrices, it is common to refer to numerical quantities as *scalars*. In this text *all our scalars will be real numbers*.

If A is a matrix, we will use a_{ij} to denote the entry that occurs in row i and column j of A. Thus, a general 3×4 matrix can be written

$$A = \begin{bmatrix} a_{11} & a_{12} & a_{13} & a_{14} \\ a_{21} & a_{22} & a_{23} & a_{24} \\ a_{31} & a_{32} & a_{33} & a_{34} \end{bmatrix}$$

Naturally, if we use B to denote the matrix, then we will use b_{ij} for the entry row i and column j. Thus a general $m \times n$ matrix might be written

$$B = \begin{bmatrix} b_{11} & b_{12} & \cdots & b_{1n} \\ b_{21} & b_{22} & \cdots & b_{2n} \\ \vdots & \vdots & & \vdots \\ b_{m1} & b_{m2} & \cdots & b_{mn} \end{bmatrix}$$

A matrix A with n rows and n columns is called a *square matrix of order n,* and the entries a_{11}, a_{22}, ..., a_{nn} are said to be on the *main diagonal* of A (see Figure 1).

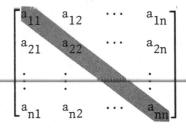

Figure 1

So far we have used matrices to abbreviate the work in solving systems of linear equations. For other applications, however, it is desirable to develop an "arithmetic of matrices" in which matrices can be added and multiplied in a useful way. The remainder of this section will be devoted to developing this arithmetic.

Two matrices are said to be *equal* if they have the same size and the corresponding entries in the two matrices are equal.

Example 2

Consider the matrices

$$A = \begin{bmatrix} 2 & 1 \\ 3 & 4 \end{bmatrix} \qquad B = \begin{bmatrix} 2 & 1 \\ 3 & 5 \end{bmatrix} \qquad C = \begin{bmatrix} 2 & 1 & 0 \\ 3 & 4 & 0 \end{bmatrix}$$

Here $A \neq C$ since A and C do not have the same size. For the same reason $B \neq C$. Also, $A \neq B$ since not all the corresponding entries are equal.

DEFINITION. If A and B are any two matrices of the same size, then the *sum* $A + B$ is the matrix obtained by adding together the corresponding entries in the two matrices. Matrices of different sizes cannot be added.

Example 3

Consider the matrices

$$A = \begin{bmatrix} 2 & 1 & 0 & 3 \\ -1 & 0 & 2 & 4 \\ 4 & -2 & 7 & 0 \end{bmatrix} \qquad B = \begin{bmatrix} -4 & 3 & 5 & 1 \\ 2 & 2 & 0 & -1 \\ 3 & 2 & -4 & 5 \end{bmatrix} \qquad C = \begin{bmatrix} 1 & 1 \\ 2 & 2 \end{bmatrix}$$

Then

$$A + B = \begin{bmatrix} -2 & 4 & 5 & 4 \\ 1 & 2 & 2 & 3 \\ 7 & 0 & 3 & 5 \end{bmatrix}$$

while $A + C$ and $B + C$ are undefined.

DEFINITION. If A is any matrix and c is any scalar, then the *product* cA is the matrix obtained by multiplying each entry of A by c.

Example 4

If A is the matrix

$$A = \begin{bmatrix} 4 & 2 \\ 1 & 3 \\ -1 & 0 \end{bmatrix}$$

then

$$2A = \begin{bmatrix} 8 & 4 \\ 2 & 6 \\ -2 & 0 \end{bmatrix} \quad \text{and} \quad (-1)A = \begin{bmatrix} -4 & -2 \\ -1 & -3 \\ 1 & 0 \end{bmatrix}$$

If B is any matrix, then -B will denote the product (-1)B. If A and B are two matrices of the same size, then A - B is defined to be the sum A + (-B) = A + (-1)B.

Example 5

Consider the matrices

$$A = \begin{bmatrix} 2 & 3 & 4 \\ 1 & 2 & 1 \end{bmatrix} \quad \text{and} \quad B = \begin{bmatrix} 0 & 2 & 7 \\ 1 & -3 & 5 \end{bmatrix}$$

From the above definitions

$$-B = \begin{bmatrix} 0 & -2 & -7 \\ -1 & 3 & -5 \end{bmatrix}$$

and

$$A - B = \begin{bmatrix} 2 & 3 & 4 \\ 1 & 2 & 1 \end{bmatrix} + \begin{bmatrix} 0 & 2 & -7 \\ -1 & 3 & -5 \end{bmatrix} = \begin{bmatrix} 2 & 1 & -3 \\ 0 & 5 & -4 \end{bmatrix}$$

Observe that A - B can be obtained directly by subtracting the entries of B from the corresponding entries of A.

Above, we defined the multiplication of a matrix by a scalar. Now we consider how two matrices are multiplied. Perhaps the most natural definition of matrix multiplication would seem to be: "multiply corresponding entries together." Surprisingly, however, this definition would not be very useful for most problems. Experience has led mathematicians to the following less intuitive but more useful definition of matrix multiplication.

DEFINITION. If A is an m × r matrix and B is an r × n matrix, then the *product* AB is the m × n matrix whose entries are determined as follows. To find the entry in row i and column j of AB, single out row i from the matrix A and column j from the matrix B. Multiply the corresponding entries from the row and column together and then add up the resulting products.

Example 6

Consider the matrices

$$A = \begin{bmatrix} 1 & 2 & 4 \\ 2 & 6 & 0 \end{bmatrix} \qquad B = \begin{bmatrix} 4 & 1 & 4 & 3 \\ 0 & -1 & 3 & 1 \\ 2 & 7 & 5 & 2 \end{bmatrix}$$

Since A is a 2 × 3 matrix and B is a 3 × 4 matrix, the product AB is a 2 × 4 matrix. To determine, for example, the entry in row 2 and column 3 of AB, we single out row 2 from A and column 3 from B. Then, as illustrated below, we multiply corresponding entries together and add up these products.

$$\begin{bmatrix} 1 & 2 & 4 \\ 2 & 6 & 0 \end{bmatrix} \begin{bmatrix} 4 & 1 & 4 & 3 \\ 0 & -1 & 3 & 1 \\ 2 & 7 & 5 & 2 \end{bmatrix} = \begin{bmatrix} \square & \square & \square & \square \\ \square & \square & \boxed{26} & \square \end{bmatrix}$$

$$(2 \cdot 4) + (6 \cdot 3) + (0 \cdot 5) = 26$$

The entry in row 1 and column 4 of AB is computed as follows.

$$\begin{bmatrix} 1 & 2 & 4 \\ 2 & 6 & 0 \end{bmatrix} \begin{bmatrix} 4 & 1 & 4 & 3 \\ 0 & -1 & 3 & 1 \\ 2 & 7 & 5 & 2 \end{bmatrix} = \begin{bmatrix} \square & \square & \square & \boxed{13} \\ \square & \square & \boxed{26} & \square \end{bmatrix}$$

$$(1 \cdot 3) + (2 \cdot 1) + (4 \cdot 2) = 13$$

The computations for the remaining products are

$(1 \cdot 4) + (2 \cdot 0) + (4 \cdot 2) = 12$

$(1 \cdot 1) - (2 \cdot 1) + (4 \cdot 7) = 27$

$(1 \cdot 4) + (2 \cdot 3) + (4 \cdot 5) = 30$

$(2 \cdot 4) + (6 \cdot 0) + (0 \cdot 2) = 8$

$(2 \cdot 1) - (6 \cdot 1) + (0 \cdot 7) = -4$

$(2 \cdot 3) + (6 \cdot 1) + (0 \cdot 2) = 12$

$$AB = \begin{bmatrix} 12 & 27 & 30 & 13 \\ 8 & -4 & 26 & 12 \end{bmatrix}$$

The definition of matrix multiplication requires that the number of columns of the first factor A be the same as the number of rows of the second factor B in order to form the product AB. If this condition is not satisfied, the product is undefined. A convenient way to determine whether a product of two matrices is defined is to write down the size of the first factor and to the right of it, write down the size of the second factor. If, as in Figure 2, the inside numbers are the same, then the product is defined. The outside numbers then give the size of the product.

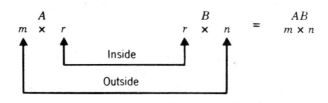

Figure 2

Example 7

Suppose that A is a 3×4 matrix, B is a 4×7 matrix, and C is a 7×3 matrix. Then AB is defined and is a 3×7 matrix; CA is defined and is a 7×4 matrix; BC is defined and is a 4×3 matrix. The products AC, CB, and BA are all undefined.

Example 8

If A is a general $m \times r$ matrix and B is a general $r \times n$ matrix, then, as suggested by the shading below, the entry in row i and column j of AB is given by the formula

$$a_{i1}b_{1j} + a_{i2}b_{2j} + a_{i3}b_{3j} + \cdots + a_{ir}b_{rj}$$

$$AB = \begin{bmatrix} a_{11} & a_{12} & \cdots & a_{1r} \\ a_{21} & a_{22} & \cdots & a_{2r} \\ \vdots & \vdots & & \vdots \\ a_{i1} & a_{i2} & \cdots & a_{ir} \\ \vdots & \vdots & & \vdots \\ a_{m1} & a_{m2} & \cdots & a_{mr} \end{bmatrix} \begin{bmatrix} b_{11} & b_{12} & \cdots & b_{1j} & \cdots & b_{1n} \\ b_{21} & b_{22} & \cdots & b_{2j} & \cdots & b_{2n} \\ \vdots & \vdots & & \vdots & & \vdots \\ b_{r1} & b_{r2} & \cdots & b_{rj} & \cdots & b_{rn} \end{bmatrix}$$

Matrix multiplication has an important application to systems of linear equations. Consider any system of m linear equations in n unknowns.

$$a_{11}x_1 + a_{12}x_2 + \cdots + a_{1n}x_n = b_1$$
$$a_{21}x_1 + a_{22}x_2 + \cdots + a_{2n}x_n = b_2$$
$$\vdots \qquad \vdots \qquad \qquad \vdots \qquad \vdots$$
$$a_{m1}x_1 + a_{m2}x_2 + \cdots + a_{mn}x_n = b_m$$

Since two matrices are equal if and only if their corresponding entries are equal, we can replace the m equations in this system by the single matrix equation

$$\begin{bmatrix} a_{11}x_1 + a_{12}x_2 + \cdots + a_{1n}x_n \\ a_{21}x_1 + a_{22}x_2 + \cdots + a_{2n}x_n \\ \vdots \qquad \vdots \qquad \qquad \vdots \\ a_{m1}x_1 + a_{m2}x_2 + \cdots + a_{mn}x_n \end{bmatrix} = \begin{bmatrix} b_1 \\ b_2 \\ \vdots \\ b_m \end{bmatrix}$$

The m × 1 matrix on the left side of this equation can be written as a product to give

$$\begin{bmatrix} a_{11} & a_{12} & \cdots & a_{1n} \\ a_{21} & a_{22} & \cdots & a_{2n} \\ \vdots & \vdots & & \vdots \\ a_{m1} & a_{m2} & \cdots & a_{mn} \end{bmatrix} \begin{bmatrix} x_1 \\ x_2 \\ \vdots \\ x_n \end{bmatrix} = \begin{bmatrix} b_1 \\ b_2 \\ \vdots \\ b_m \end{bmatrix}$$

If we designate these matrices by A, X, and B, respectively, the original system of m equations in n unknowns has been replaced by the single matrix equation

$$AX = B \qquad\qquad (1)$$

Some of our later work will be devoted to solving matrix equations like this for the unknown matrix X. As a consequence of this matrix approach, we will obtain effective new methods for solving systems of linear equations. The matrix A in (1) is called the *coefficient matrix* for the system.

Example 9

At times it is helpful to be able to find a particular row or column in a product AB without computing the entire product. We leave it as an exercise to show that the entries in the jth column of AB are the entries in the product AB_j, where B_j is the matrix formed using only the jth column of B. Thus, if A and B are the matrices in Example 6, the second column of AB can be obtained by the computation

$$\begin{bmatrix} 1 & 2 & 4 \\ 2 & 6 & 0 \end{bmatrix} \begin{bmatrix} 1 \\ -1 \\ 7 \end{bmatrix} = \begin{bmatrix} 27 \\ -4 \end{bmatrix}$$

$$\uparrow \qquad\qquad \uparrow$$

second column   second column
   of B   of AB

Similarly, the entries in the ith row of AB are the entries in the product A_iB, where A_i is the matrix formed by using only the ith row of A. Thus, the first row in the product AB of Example 6 can be obtained by the computation

$$[1 \quad 2 \quad 4] \begin{bmatrix} 4 & 1 & 4 & 3 \\ 0 & -1 & 3 & 1 \\ 2 & 7 & 5 & 2 \end{bmatrix} = [12 \quad 27 \quad 30 \quad 13]$$

first row of A   first row of AB

We conclude this section with an important matrix operation that has no analog in ordinary arithmetic.

If A is any $m \times n$ matrix, then the *transpose of A* is denoted by A^t and is defined to be the $n \times m$ matrix whose first column is the first row of A, whose second column is the second row of A, whose third column is the third row of A, etc.

Example 10

The transposes of the matrices

$$A = \begin{bmatrix} a_{11} & a_{12} & a_{13} & a_{14} \\ a_{21} & a_{22} & a_{23} & a_{24} \\ a_{31} & a_{32} & a_{33} & a_{34} \end{bmatrix}$$

$$B = \begin{bmatrix} 2 & 3 \\ 1 & 4 \\ 5 & 6 \end{bmatrix} \quad C = \begin{bmatrix} 1 & 3 & 5 \end{bmatrix} \quad D = \begin{bmatrix} 3 & 5 & -2 \\ 5 & 4 & 1 \\ -2 & 1 & 7 \end{bmatrix}$$

are

$$A^t = \begin{bmatrix} a_{11} & a_{21} & a_{31} \\ a_{12} & a_{22} & a_{32} \\ a_{13} & a_{23} & a_{33} \\ a_{14} & a_{24} & a_{34} \end{bmatrix}$$

$$B^t = \begin{bmatrix} 2 & 1 & 5 \\ 3 & 4 & 6 \end{bmatrix} \quad C^t = \begin{bmatrix} 1 \\ 3 \\ 5 \end{bmatrix} \quad D^t = \begin{bmatrix} 3 & 5 & -2 \\ 5 & 4 & 1 \\ -2 & 1 & 7 \end{bmatrix}$$

EXERCISE SET 1.4

1. Let A and B be 4×5 matrices and let C, D, and E be 5×2, 4×2, and 5×4 matrices, respectively. Determine which of the following matrix expressions are defined. For those that are defined, give the size of the resulting matrix.

(a) BA   (b) AC + D   (c) AE + B

(d) AB + B   (e) E(A + B)   (f) E(AC)

2. (a) Show that if AB and BA are both defined, then AB and BA are square matrices.

   (b) Show that if A is an $m \times n$ matrix and A(BA) is defined, then B is an $n \times m$ matrix.

3. Solve the following matrix equation for a, b, c, and d.

$$\begin{bmatrix} a-b & b+c \\ 3d+c & 2a-4d \end{bmatrix} = \begin{bmatrix} 8 & 1 \\ 7 & 6 \end{bmatrix}$$

4. Consider the matrices

$$A = \begin{bmatrix} 3 & 0 \\ -1 & 2 \\ 1 & 1 \end{bmatrix} \qquad B = \begin{bmatrix} 4 & -1 \\ 0 & 2 \end{bmatrix} \qquad C = \begin{bmatrix} 1 & 4 & 2 \\ 3 & 1 & 5 \end{bmatrix}$$

$$D = \begin{bmatrix} 1 & 5 & 2 \\ -1 & 0 & 1 \\ 3 & 2 & 4 \end{bmatrix} \qquad E = \begin{bmatrix} 6 & 1 & 3 \\ -1 & 1 & 2 \\ 4 & 1 & 3 \end{bmatrix}$$

Compute

(a) AB   (b) D + E   (c) D - E   (d) DE

(e) ED   (f) -7B   (g) $A^t + C$ (h) $2D^t - E^t$.

5. Using the matrices from Exercise 4, compute (where possible)

(a) 3C - D   (b) (3E)D   (c) (AB)C

(d) A(BC)   (e) (4B)C + 2B   (f) $D + E^2$ (where $E^2 = EE$)

(g) $A^t B$ (h) $C^t B$ (i) $(AB)^t$

6. Let

$$A = \begin{bmatrix} 3 & -2 & 7 \\ 6 & 5 & 4 \\ 0 & 4 & 9 \end{bmatrix} \quad \text{and} \quad B = \begin{bmatrix} 6 & -2 & 4 \\ 0 & 1 & 3 \\ 7 & 7 & 5 \end{bmatrix}$$

Use the method of Example 9 to find

(a) the first row of AB   (b) the third row of AB

(c) the second column of AB   (d) the first column of BA

(e) the third row of AA   (f) the third column of AA.

7. Let C, D, and E be the matrices in Exercise 4. Using as few computations as possible, determine the entry in row 2 and column 3 of C(DE).

8. (a) Show that if A has a row of zeros and B is any matrix for which AB is defined, then AB also has a row of zeros.

   (b) Find a similar result involving a column of zeros.

9. Let A be any $m \times n$ matrix and let 0 be the $m \times n$ matrix each of whose entries is zero. Show that if kA = 0, then k = 0 or A = 0.

10. Let I be the $n \times n$ matrix whose entry in row i and column j is

$$\begin{cases} 1 & \text{if} \quad i = j \\ 0 & \text{if} \quad i \neq j \end{cases}$$

Show that AI = IA = A for every $n \times n$ matrix A.

11. A square matrix is called a *diagonal matrix* if all entries off the main diagonal are zeros. Show that the product of diagonal matrices is again a diagonal matrix. State a rule for multiplying diagonal matrices.

12. (a) Show that the entries in the jth column of AB are the entries in the product AB_j, where B_j is the matrix formed from the jth column of B.

   (b) Show that the entries in the ith row of AB are the entries in the product A_iB, where A_i is the matrix formed from the ith row of A.

1.5 RULES OF MATRIX ARITHMETIC

Although many of the rules of arithmetic for real numbers also hold for matrices, there are some exceptions. One of the most important exceptions occurs in the multiplication of matrices. For real numbers a and b, we always have ab = ba. This property is often called the *commutative law for multiplication*. For matrices, however, AB and BA need not be equal. Equality can fail to hold for three reasons. It can happen, for example, that AB is defined but BA is undefined. This is the case if A is a 2×3 matrix and B is a 3×4 matrix. Also, it can happen that AB and BA are both defined but have different sizes. This is the situation if A is a 2×3 matrix and B is a 3×2 matrix. Finally, as our next example shows, it is possible to have AB ≠ BA even if both AB and BA are defined and have the same size.

Example 1

Consider the matrices

$$A = \begin{bmatrix} -1 & 0 \\ 2 & 3 \end{bmatrix} \qquad B = \begin{bmatrix} 1 & 2 \\ 3 & 0 \end{bmatrix}$$

Multiplying gives

$$AB = \begin{bmatrix} -1 & -2 \\ 11 & 4 \end{bmatrix} \qquad BA = \begin{bmatrix} 3 & 6 \\ -3 & 0 \end{bmatrix}$$

Thus AB ≠ BA.

Although the commutative law for multiplication is not valid in matrix arithmetic, many familiar laws of arithmetic are valid for matrices. Some of the most important ones and their names are summarized in the following theorem.

THEOREM 1. *Assuming that the sizes of the matrices are such that the indicated operations can be performed, the following rules of matrix arithmetic are valid.*

(a) A + B = B + A   *(Commutative law for addition)*
(b) A + (B + C) = (A + B) + C   *(Associative law for addition)*
(c) A(BC) = (AB)C   *(Associative law for multiplication)*
(d) A(B + C) = AB + AC   *(Distributive law)*
(e) (B + C)A = BA + CA   *(Distributive law)*
(f) A(B - C) = AB - AC
(g) (B - C)A = BA - CA
(h) a(B + C) = aB + aC

(i) $a(B - C) = aB - aC$
(j) $(a + b)C = aC + bC$
(k) $(a - b)C = aC - bC$
(l) $(ab)C = a(bC)$
(m) $a(BC) = (aB)C = B(aC)$

Each of the equations in this theorem asserts an equality between matrices. To prove one of these equalities, it is necessary to show that the matrix on the left side has the same size as the matrix on the right side, and that corresponding entries on the two sides are equal.

Although the operations of matrix addition and matrix multiplication were defined for pairs of matrices, associative laws (b) and (c) enable us to denote sums and products of three matrices as $A + B + C$ and ABC without inserting any parentheses. This is justified by the fact that no matter how parentheses are inserted, the associative laws guarantee that the same end result will be obtained. Without going into details we observe that similar results are valid for sums or products involving four or more matrices. In general, *given any sum or any product of matrices, pairs of parentheses can be inserted or deleted anywhere within the expression without affecting the end result.*

Example 2

As an illustration of the associative law for matrix multiplication, consider

$$A = \begin{bmatrix} 1 & 2 \\ 3 & 4 \\ 0 & 1 \end{bmatrix} \qquad B = \begin{bmatrix} 4 & 3 \\ 2 & 1 \end{bmatrix} \qquad C = \begin{bmatrix} 1 & 0 \\ 2 & 3 \end{bmatrix}$$

Then

$$AB = \begin{bmatrix} 1 & 2 \\ 3 & 4 \\ 0 & 1 \end{bmatrix} \begin{bmatrix} 4 & 3 \\ 2 & 1 \end{bmatrix} = \begin{bmatrix} 8 & 5 \\ 20 & 13 \\ 2 & 1 \end{bmatrix}$$

so that

$$(AB)C = \begin{bmatrix} 8 & 5 \\ 20 & 13 \\ 2 & 1 \end{bmatrix} \begin{bmatrix} 1 & 0 \\ 2 & 3 \end{bmatrix} = \begin{bmatrix} 18 & 15 \\ 46 & 39 \\ 4 & 3 \end{bmatrix}$$

On the other hand

$$BC = \begin{bmatrix} 4 & 3 \\ 2 & 1 \end{bmatrix} \begin{bmatrix} 1 & 0 \\ 2 & 3 \end{bmatrix} = \begin{bmatrix} 10 & 9 \\ 4 & 3 \end{bmatrix}$$

so that

$$A(BC) = \begin{bmatrix} 1 & 2 \\ 3 & 4 \\ 0 & 1 \end{bmatrix} \begin{bmatrix} 10 & 9 \\ 4 & 3 \end{bmatrix} = \begin{bmatrix} 18 & 15 \\ 46 & 39 \\ 4 & 3 \end{bmatrix}$$

Thus $(AB)C = A(BC)$, as guaranteed by Theorem 1(c).

A matrix, all of whose entries are zero, such as

$$\begin{bmatrix} 0 & 0 \\ 0 & 0 \end{bmatrix}, \quad \begin{bmatrix} 0 & 0 & 0 \\ 0 & 0 & 0 \\ 0 & 0 & 0 \end{bmatrix}, \quad \begin{bmatrix} 0 & 0 & 0 & 0 \\ 0 & 0 & 0 & 0 \end{bmatrix}, \quad \begin{bmatrix} 0 \\ 0 \\ 0 \\ 0 \end{bmatrix}, \quad [0]$$

is called a *zero matrix*. Zero matrices will be denoted by 0; if it is important to emphasize the size, we shall write $0_{m \times n}$ for the $m \times n$ zero matrix.

If A is any matrix and 0 is the zero matrix with the same size, it is obvious that $A + 0 = A$. The matrix 0 plays much the same role in this matrix equation as the number 0 plays in the numerical equation $a + 0 = a$.

Since we already know that some of the rules of arithmetic for real numbers do not carry over to matrix arithmetic, it would be foolhardy to assume that all the properties of the real number zero carry over to zero matrices. For example, consider the following two standard results in the arithmetic of real numbers.

(i) If $ab = ac$ and $a \neq 0$, then $b = c$. (This is called the *cancellation law*.)
(ii) If $ad = 0$, then at least one of the factors on the left is 0.

As the next example shows, the corresponding results are false in matrix arithmetic.

Example 3

Consider the matrices

$$A = \begin{bmatrix} 0 & 1 \\ 0 & 2 \end{bmatrix} \qquad B = \begin{bmatrix} 1 & 1 \\ 3 & 4 \end{bmatrix} \qquad C = \begin{bmatrix} 2 & 5 \\ 3 & 4 \end{bmatrix} \qquad D = \begin{bmatrix} 3 & 7 \\ 0 & 0 \end{bmatrix}$$

Here

$$AB = AC = \begin{bmatrix} 3 & 4 \\ 6 & 8 \end{bmatrix}$$

Although $A \neq 0$, it is *incorrect* to cancel the A from both sides of the equation $AB = AC$ and write $B = C$. Thus the cancellation law fails to hold for matrices.

Also, $AD = 0$; yet $A \neq 0$ and $D \neq 0$ so that result (ii) listed above does not carry over to matrix arithmetic.

In spite of these negative examples, a number of familiar properties of the real number 0 carry over to zero matrices. Some of the more important ones are summarized in the next theorem.

THEOREM 2. *Assuming that the sizes of the matrices are such that the indicated operations can be performed, the following rules of matrix arithmetic are valid.*

(a) $A + 0 = 0 + A = A$
(b) $A - A = 0$
(c) $0 - A = -A$
(d) $A0 = 0; \quad 0A = 0$

As an application of our results on matrix arithmetic, we prove the following theorem, which was anticipated earlier in the text.

THEOREM 3. *Every system of linear equations has either no solutions, exactly one solution, or infinitely many solutions.*

Proof. If $AX = B$ is a system of linear equations, exactly one of the following is true: (a) the system has no solutions, (b) the system has exactly one solution, or (c) the system has more than one solution. The proof will be complete if we can show that the system has infinitely many solutions in case (c).

Assume that $AX = B$ has more than one solution and let X_1 and X_2 be two different solutions. Thus, $AX_1 = B$ and $AX_2 = B$. Subtracting these equations gives $AX_1 - AX_2 = 0$ or $A(X_1 - X_2) = 0$. If we let $X_0 = X_1 - X_2$ and let k be any scalar, then

$$A(X_1 + kX_0) = AX_1 + A(kX_0)$$
$$= AX_1 + k(AX_0)$$
$$= B + k0$$
$$= B + 0$$
$$= B$$

But this says that $X_1 + kX_0$ is a solution of $AX = B$. Since there are infinitely many choices for k, $AX = B$ has infinitely many solutions. ☐

Of special interest are square matrices with 1's on the main diagonal and 0's off the main diagonal, such as

$$\begin{bmatrix} 1 & 0 \\ 0 & 1 \end{bmatrix}, \quad \begin{bmatrix} 1 & 0 & 0 \\ 0 & 1 & 0 \\ 0 & 0 & 1 \end{bmatrix}, \quad \begin{bmatrix} 1 & 0 & 0 & 0 \\ 0 & 1 & 0 & 0 \\ 0 & 0 & 1 & 0 \\ 0 & 0 & 0 & 1 \end{bmatrix}, \quad \text{etc.}$$

A matrix of this kind is called an *identity matrix* and is denoted by I. If it is important to emphasize the size, we shall write I_n for the $n \times n$ identity matrix.

If A is an $m \times n$ matrix, then, as illustrated in the next example, $AI_n = A$ and $I_m A = A$. Thus, an identity matrix plays much the same role in matrix arithmetic as the number 1 plays in the numerical relationships $a \cdot 1 = 1 \cdot a = a$.

Example 4

Consider the matrix

$$A = \begin{bmatrix} a_{11} & a_{12} & a_{13} \\ a_{21} & a_{22} & a_{23} \end{bmatrix}$$

Then

$$I_2 A = \begin{bmatrix} 1 & 0 \\ 0 & 1 \end{bmatrix} \begin{bmatrix} a_{11} & a_{12} & a_{13} \\ a_{21} & a_{22} & a_{23} \end{bmatrix} = \begin{bmatrix} a_{11} & a_{12} & a_{13} \\ a_{21} & a_{22} & a_{23} \end{bmatrix} = A$$

and

$$AI_3 = \begin{bmatrix} a_{11} & a_{12} & a_{13} \\ a_{21} & a_{22} & a_{23} \end{bmatrix} \begin{bmatrix} 1 & 0 & 0 \\ 0 & 1 & 0 \\ 0 & 0 & 1 \end{bmatrix} = \begin{bmatrix} a_{11} & a_{12} & a_{13} \\ a_{21} & a_{22} & a_{23} \end{bmatrix} = A$$

If A is any square matrix, and if a matrix B can be found such that AB = BA = I, then A is said to be *invertible* and B is called an *inverse* of A.

Example 5

The matrix

$$B = \begin{bmatrix} 3 & 5 \\ 1 & 2 \end{bmatrix} \qquad \text{is an inverse of} \qquad A = \begin{bmatrix} 2 & -5 \\ -1 & 3 \end{bmatrix}$$

since

$$AB = \begin{bmatrix} 2 & -5 \\ -1 & 3 \end{bmatrix} \begin{bmatrix} 3 & 5 \\ 1 & 2 \end{bmatrix} = \begin{bmatrix} 1 & 0 \\ 0 & 1 \end{bmatrix} = I$$

and

$$BA = \begin{bmatrix} 3 & 5 \\ 1 & 2 \end{bmatrix} \begin{bmatrix} 2 & -5 \\ -1 & 3 \end{bmatrix} = \begin{bmatrix} 1 & 0 \\ 0 & 1 \end{bmatrix} = I$$

Example 6

The matrix

$$A = \begin{bmatrix} 1 & 4 & 0 \\ 2 & 5 & 0 \\ 3 & 6 & 0 \end{bmatrix}$$

is not invertible. To see why, let

$$B = \begin{bmatrix} b_{11} & b_{12} & b_{13} \\ b_{21} & b_{22} & b_{23} \\ b_{31} & b_{32} & b_{33} \end{bmatrix}$$

by any 3×3 matrix. From Example 9 in Section 1.4 the third column of BA is

$$\begin{bmatrix} b_{11} & b_{12} & b_{13} \\ b_{21} & b_{22} & b_{23} \\ b_{31} & b_{32} & b_{33} \end{bmatrix} \begin{bmatrix} 0 \\ 0 \\ 0 \end{bmatrix} = \begin{bmatrix} 0 \\ 0 \\ 0 \end{bmatrix}$$

Thus

$$BA \neq I = \begin{bmatrix} 1 & 0 & 0 \\ 0 & 1 & 0 \\ 0 & 0 & 1 \end{bmatrix}$$

It is reasonable to ask whether an invertible matrix can have more than one inverse. The next theorem shows the answer is no—an invertible matrix has exactly one inverse.

THEOREM 4. *If B and C are both inverses of the matrix A, then B = C.*

Proof. Since B is an inverse of A, BA = I. Multiplying both sides on the right by C gives (BA)C = IC = C. But (BA)C = B(AC) = BI = B, so that B = C. □

As a consequence of this important result, we can now speak of "the" inverse of an invertible matrix. If A is invertible, then its inverse will be denoted by the symbol A^{-1}. Thus

$$AA^{-1} = I \qquad \text{and} \qquad A^{-1}A = I$$

The inverse of A plays much the same role in matrix arithmetic that reciprocal a^{-1} plays in the numerical relationships $aa^{-1} = 1$ and $a^{-1}a = 1$.

Example 7

Consider the 2×2 matrix

$$A = \begin{bmatrix} a & b \\ c & d \end{bmatrix}$$

If $ad - bc \neq 0$, then

$$A^{-1} = \frac{1}{ad - bc} \begin{bmatrix} d & -b \\ -c & a \end{bmatrix} = \begin{bmatrix} \dfrac{d}{ad - bc} & -\dfrac{b}{ad - bc} \\ -\dfrac{c}{ad - bc} & \dfrac{a}{ad - bc} \end{bmatrix}$$

since $AA^{-1} = I_2$ and $A^{-1}A = I_2$ (verify). In the next section we shall show how to find inverses of invertible matrices whose sizes are greater than 2×2.

THEOREM 5.   *If A and B are invertible matrices of the same size, then*

(a) AB is invertible
(b) $(AB)^{-1} = B^{-1}A^{-1}$

Proof. If we can show that $(AB)(B^{-1}A^{-1}) = (B^{-1}A^{-1})(AB) = I$, then we will have simultaneously established that AB is invertible and that $(AB)^{-1} = B^{-1}A^{-1}$. But $(AB)(B^{-1}A^{-1}) = A(BB^{-1})A^{-1} = AIA^{-1} = AA^{-1} = I$. Similarly $(B^{-1}A^{-1})(AB) = I$. □

Although we will not prove it, this result can be extended to include three or more factors. Thus, we can state the following general rule:

A product of invertible matrices is always invertible, and the inverse of the product is the product of the inverses in the reverse order.

Example 8

Consider the matrices

$$A = \begin{bmatrix} 1 & 2 \\ 1 & 3 \end{bmatrix} \qquad B = \begin{bmatrix} 3 & 2 \\ 2 & 2 \end{bmatrix} \qquad AB = \begin{bmatrix} 7 & 6 \\ 9 & 8 \end{bmatrix}$$

Applying the formula given in Example 7, we obtain

$$A^{-1} = \begin{bmatrix} 3 & -2 \\ -1 & 1 \end{bmatrix} \qquad B^{-1} = \begin{bmatrix} 1 & -1 \\ -1 & \frac{3}{2} \end{bmatrix} \qquad (AB)^{-1} = \begin{bmatrix} 4 & -3 \\ -\frac{9}{2} & \frac{7}{2} \end{bmatrix}$$

Also

$$B^{-1}A^{-1} = \begin{bmatrix} 1 & -1 \\ -1 & \frac{3}{2} \end{bmatrix} \begin{bmatrix} 3 & -2 \\ -1 & 1 \end{bmatrix} = \begin{bmatrix} 4 & -3 \\ -\frac{9}{2} & \frac{7}{2} \end{bmatrix}$$

Therefore $(AB)^{-1} = B^{-1}A^{-1}$ as guaranteed by Theorem 5.

If A is a square matrix and n is a positive integer, we define

$$A^n = \underbrace{AA\cdots A}_{n \text{ factors}}$$

$$A^0 = I$$

If, in addition, A is invertible, we define

$$A^{-n} = (A^{-1})^n = \underbrace{A^{-1}A^{-1}\cdots A^{-1}}_{n \text{ factors}}$$

The following theorem shows that the familiar laws of exponents are valid for matrices.

THEOREM 6. *If A is a square matrix and r and s are integers then*

$$A^r A^s = A^{r+s} \qquad (A^r)^s = A^{rs}$$

We conclude this section with two theorems that list some additional properties of matrix operations.

THEOREM 7. *If A is an invertible matrix, then:*

(a) A^{-1} *is invertible and* $(A^{-1})^{-1} = A$.

(b) A^n *is invertible and* $(A^n)^{-1} = (A^{-1})^n$ *for* $n = 0, 1, 2, \ldots$.

(c) *For any nonzero scalar* k, kA *is invertible and* $(kA)^{-1} = \frac{1}{k} A^{-1}$.

THEOREM 8. *If the sizes of the matrices are such that operations can be performed, then*

(a) $(A^t)^t = A$

(b) $(A + B)^t = A^t + B^t$

(c) $(kA)^t = kA^t$, *where* k *is any scalar*

(d) $(AB)^t = B^t A^t$

The last result in this theorem can be extended to include three or more factors. Thus, we have the following rule:

> The transpose of a product of matrices is the product of their transposes in the reverse order.

EXERCISE SET 1.5

1. Let

$$A = \begin{bmatrix} 3 & 2 \\ -1 & 3 \end{bmatrix} \quad B = \begin{bmatrix} 4 & 0 \\ 1 & 5 \end{bmatrix} \quad C = \begin{bmatrix} 0 & -1 \\ 4 & 6 \end{bmatrix} \quad a = -3 \quad b = 2 .$$

Show

(a) $A + (B + C) = (A + B) + C$

(b) $(AB)C = A(BC)$

(c) $(a + b)C = aC + bC$

(d) $a(B - C) = aB - aC$.

2. Using the matrices and scalars in Exercise 1, show

(a) $a(BC) = (aB)C = B(aC)$

(b) $A(B - C) = AB - AC$.

3. Use the formula given in Example 7 to compute the inverses of the following matrices.

$$A = \begin{bmatrix} 3 & 1 \\ 5 & 2 \end{bmatrix} \quad B = \begin{bmatrix} 2 & -3 \\ 4 & 4 \end{bmatrix} \quad C = \begin{bmatrix} 2 & 0 \\ 0 & 3 \end{bmatrix}$$

4. Verify that the matrices A and B in Exercise 3 satisfy the relationship $(AB)^{-1} = B^{-1}A^{-1}$.

5. Let A and B be square matrices of the same size. Is $(AB)^2 = A^2 B^2$ a valid matrix identity? Justify your answer.

6. Let A be an invertible matrix whose inverse is

$$\begin{bmatrix} 3 & 4 \\ 5 & 6 \end{bmatrix}$$

Find the matrix A.

7. Let A be an invertible matrix, and suppose that the inverse of 7A is

$$\begin{bmatrix} -1 & 2 \\ 4 & -7 \end{bmatrix}$$

Find the matrix A.

8. Let A be the matrix

$$\begin{bmatrix} 1 & 0 \\ 2 & 3 \end{bmatrix}$$

Compute A^3, A^{-3}, and $A^2 - 2A + I$.

9. Let A be the matrix

$$\begin{bmatrix} 1 & 1 & 0 \\ 0 & 1 & 1 \\ 1 & 0 & 1 \end{bmatrix}$$

Determine whether A is invertible, and if so, find its inverse.
(*Hint.* Solve AX = I by equating corresponding entries on the two sides.)

10. Find the inverse of

$$\begin{bmatrix} \cos\theta & \sin\theta \\ -\sin\theta & \cos\theta \end{bmatrix}$$

11. (a) Find 2×2 matrices A and B such that

$$(A + B)^2 \neq A^2 + 2AB + B^2$$

(b) Show that, if A and B are square matrices such that AB = BA, then

$$(A + B)^2 = A^2 + 2AB + B^2$$

(c) Find an expansion of $(A + B)^2$ that is valid for all square matrices A and B having the same size.

12. Consider the matrix

$$
A = \begin{bmatrix}
a_{11} & 0 & 0 & \cdots & 0 \\
0 & a_{22} & 0 & \cdots & 0 \\
\vdots & \vdots & \vdots & & \vdots \\
0 & 0 & 0 & \cdots & a_{nn}
\end{bmatrix}
$$

where $a_{11} a_{22} \cdots a_{nn} \neq 0$. Show that A is invertible and find its inverse.

13. Let

$$
A = \begin{bmatrix}
a_{11} & a_{12} & a_{13} \\
a_{21} & a_{22} & a_{23} \\
a_{31} & a_{32} & a_{33}
\end{bmatrix}
\quad \text{and} \quad
B = \begin{bmatrix}
b_{11} & b_{12} & b_{13} \\
b_{21} & b_{22} & b_{23} \\
b_{31} & b_{32} & b_{33}
\end{bmatrix}
$$

Show that

(a) $(A^t)^t = A$ (b) $(A + B)^t = A^t + B^t$ (c) $(AB)^t = B^t A^t$ (d) $(kA)^t = kA^t$

14. (a) Find a nonzero 3×3 matrix A such that $A = A^t$.

(b) Find a nonzero 3×3 matrix A such that $A = -A^t$.

15. Let a_{ij} be the entry in the ith row and jth column of A. In which row and column of A^t will a_{ij} appear?

16. A square matrix A is called *symmetric* if $A^t = A$ and *skew-symmetric* if $A^t = -A$. Show that if B is a square matrix, then

(a) BB^t and $B + B^t$ are symmetric   (b) $B - B^t$ is skew-symmetric

17. Assume that A is a square matrix which satisfies $A^2 - 3A + I = 0$. Show that $A^{-1} = 3I - A$.

18. (a) Show that a matrix with a row of zeros cannot have an inverse.

    (b) Show that a matrix with a column of zeros cannot have an inverse.

19. Is the sum of two invertible matrices necessarily invertible?

20. Let A and B be square matrices such that AB = 0. Show that A cannot be invertible unless B = 0.

21. In Theorem 2 why didn't we write part (d) as A0 = 0 = 0A?

22. The real equation $a^2 = 1$ has exactly two solutions. Find at least eight different 3×3 matrices that satisfy the matrix equation $A^2 = I_3$. (*Hint.* Look for solutions in which all the entries off the main diagonal are zero.)

23. Let AX = B be any consistent system of linear equations, and let X_1 be a fixed solution. Show that every solution to the system can be written in the form $X = X_1 + X_0$ where X_0 is a solution to AX = 0. Show also that every matrix of this form is a solution.

24. (a) Show that if A is invertible and AB = AC then B = C.

    (b) Explain why part (a) and Example 3 do not contradict one another.

1.6 A METHOD FOR FINDING A^{-1}

In this section we shall give an algorithm (procedure) for finding the inverse of an invertible matrix. We begin with some preliminary results.

If a matrix B can be obtained from a matrix A by performing a finite sequence of elementary row operations, then obviously we can get from B back to A by performing the inverses of these elementary row operations in reverse order. Matrices that can be obtained from one another by a finite sequence of elementary row operations are said to be *row equivalent*.

The next theorem establishes some fundamental relationships between $n \times n$ matrices and systems of n linear equations in n unknowns. These results are extremely important and will be used many times in later sections.

Theorem 1. If A is an $n \times n$ matrix, then the following statements are equivalent, that is, are all true or all false.

(a) A is invertible.

(b) AX = 0 has only the trivial solution.

(c) A is row equivalent to I_n.

This theorem tells us that if an $n \times n$ matrix A is invertible, then there is a finite sequence of row operations that reduces A to I_n. The following theorem is the heart of the algorithm for finding A^{-1}.

Theorem 2. If A is an $n \times n$ invertible matrix, then the sequence of row operations that reduces A to I_n reduces I_n to A^{-1}.

Thus, to find the inverse of an invertible matrix A, we must find a sequence of elementary row operations that reduces A to the identity and then perform this same sequence of operations on I_n to obtain A^{-1}. A simple method for carrying out this procedure is given in the following example.

Example 1

Find the inverse of

$$A = \begin{bmatrix} 1 & 2 & 3 \\ 2 & 5 & 3 \\ 1 & 0 & 8 \end{bmatrix}$$

We wish to reduce A to the identity matrix by row operations and simultaneously apply these operations to I to produce A^{-1}. This can be accomplished by adjoining the identity matrix to the right of A and applying row operations to both sides until the left side is reduced to I. The final matrix will then have the form $I|A^{-1}$. The computations can be carried out as follows.

$$\begin{bmatrix} 1 & 2 & 3 & | & 1 & 0 & 0 \\ 2 & 5 & 3 & | & 0 & 1 & 0 \\ 1 & 0 & 8 & | & 0 & 0 & 1 \end{bmatrix}$$

$$\begin{bmatrix} 1 & 2 & 3 & | & 1 & 0 & 0 \\ 0 & 1 & -3 & | & -2 & 1 & 0 \\ 0 & -2 & 5 & | & -1 & 0 & 1 \end{bmatrix}$$

We added -2 times the first row to the second and -1 times the first row to the third.

$$\begin{bmatrix} 1 & 2 & 3 & | & 1 & 0 & 0 \\ 0 & 1 & -3 & | & -2 & 1 & 0 \\ 0 & 0 & -1 & | & -5 & 2 & 1 \end{bmatrix}$$

We added 2 times the second row to the third.

$$\begin{bmatrix} 1 & 2 & 3 & | & 1 & 0 & 0 \\ 0 & 1 & -3 & | & -2 & 1 & 0 \\ 0 & 0 & 1 & | & 5 & -2 & -1 \end{bmatrix}$$

We multiplied the third row by -1.

$$\begin{bmatrix} 1 & 2 & 0 & | & -14 & 6 & 3 \\ 0 & 1 & 0 & | & 13 & -5 & -3 \\ 0 & 0 & 1 & | & 5 & -2 & -1 \end{bmatrix}$$

We added 3 times the third row to the second and -3 times the third row to the first.

$$\begin{bmatrix} 1 & 0 & 0 & | & -40 & 16 & 9 \\ 0 & 1 & 0 & | & 13 & -5 & -3 \\ 0 & 0 & 1 & | & 5 & -2 & -1 \end{bmatrix}$$

We added -2 times the second row to the first.

Thus

$$A^{-1} = \begin{bmatrix} -40 & 16 & 9 \\ 13 & -5 & -3 \\ 5 & -2 & -1 \end{bmatrix}$$

Often it will not be known in advance whether a given matrix is invertible. If the procedure used in this example is attempted on a matrix that is not invertible, then, by part (c) of Theorem 1, it will be impossible to reduce the left side to I by row operations. At some point in the computation a row of zeros will occur on the left side, and it can then be concluded that the given matrix is not invertible. The computations can then be stopped.

Example 2

Consider the matrix

$$A = \begin{bmatrix} 1 & 6 & 4 \\ 2 & 4 & -1 \\ -1 & 2 & 5 \end{bmatrix}$$

Applying the procedure of Example 1 yields

$$\left[\begin{array}{rrr|rrr} 1 & 6 & 4 & 1 & 0 & 0 \\ 2 & 4 & -1 & 0 & 1 & 0 \\ -1 & 2 & 5 & 0 & 0 & 1 \end{array}\right]$$

$$\left[\begin{array}{rrr|rrr} 1 & 6 & 4 & 1 & 0 & 0 \\ 0 & -8 & -9 & -2 & 1 & 0 \\ 0 & 8 & 9 & 1 & 0 & 1 \end{array}\right]$$

We added -2 times the first row to the second and added the first row to the third.

$$\left[\begin{array}{rrr|rrr} 1 & 6 & 4 & 1 & 0 & 0 \\ 0 & -8 & -9 & -2 & 1 & 0 \\ 0 & 0 & 0 & -1 & 1 & 1 \end{array}\right]$$

We added the second row to the third.

Since we have obtained a row of zeros on the left side, A is not invertible.

Example 3

In Example 1 we showed that

$$A = \begin{bmatrix} 1 & 2 & 3 \\ 2 & 5 & 3 \\ 1 & 0 & 8 \end{bmatrix}$$

is an invertible matrix. From Theorem 1 we can now conclude that the system of equations

$$x_1 + 2x_2 + 3x_3 = 0$$
$$2x_1 + 5x_2 + 3x_3 = 0$$
$$x_1 \qquad + 8x_3 = 0$$

has only the trivial solution.

EXERCISE SET 1.6

In Exercises 1 - 12 use the method shown in Examples 1 and 2 to find the inverse of the given matrix if the matrix is invertible.

1. $\begin{bmatrix} 1 & 2 \\ 3 & 5 \end{bmatrix}$

2. $\begin{bmatrix} -2 & 3 \\ 3 & -5 \end{bmatrix}$

3. $\begin{bmatrix} 8 & -6 \\ -4 & 3 \end{bmatrix}$

4. $\begin{bmatrix} 3 & 4 & -1 \\ 1 & 0 & 3 \\ 2 & 5 & -4 \end{bmatrix}$

5. $\begin{bmatrix} 3 & 1 & 5 \\ 2 & 4 & 1 \\ -4 & 2 & -9 \end{bmatrix}$

6. $\begin{bmatrix} 1 & 0 & 1 \\ 0 & 1 & 1 \\ 1 & 1 & 0 \end{bmatrix}$

7. $\begin{bmatrix} 2 & 6 & 6 \\ 2 & 7 & 6 \\ 2 & 7 & 7 \end{bmatrix}$

8. $\begin{bmatrix} 1 & 0 & 1 \\ -1 & 1 & 1 \\ 0 & 1 & 0 \end{bmatrix}$

9. $\begin{bmatrix} \frac{1}{5} & \frac{1}{5} & \frac{1}{5} \\ \frac{1}{5} & \frac{1}{5} & -\frac{4}{5} \\ -\frac{2}{5} & \frac{1}{10} & \frac{1}{10} \end{bmatrix}$

10. $\begin{bmatrix} \sqrt{\frac{1}{2}} & \sqrt{\frac{1}{2}} & 0 \\ -\sqrt{\frac{1}{2}} & \sqrt{\frac{1}{2}} & 0 \\ 0 & 0 & 1 \end{bmatrix}$

11. $\begin{bmatrix} 1 & 0 & 0 & 0 \\ 1 & 2 & 0 & 0 \\ 1 & 2 & 4 & 0 \\ 1 & 2 & 4 & 8 \end{bmatrix}$

12. $\begin{bmatrix} 5 & 11 & 7 & 3 \\ 2 & 1 & 4 & -5 \\ 3 & -2 & 8 & 7 \\ 0 & 0 & 0 & 0 \end{bmatrix}$

13. Show that the matrix

$$A = \begin{bmatrix} \cos\theta & \sin\theta & 0 \\ -\sin\theta & \cos\theta & 0 \\ 0 & 0 & 1 \end{bmatrix}$$

is invertible for all values of θ and find A^{-1}.

14. Find the inverse of each of the following 4×4 matrices, where k_1, k_2, k_3, k_4, and k are all nonzero.

(a) $\begin{bmatrix} k_1 & 0 & 0 & 0 \\ 0 & k_2 & 0 & 0 \\ 0 & 0 & k_3 & 0 \\ 0 & 0 & 0 & k_4 \end{bmatrix}$
(b) $\begin{bmatrix} 0 & 0 & 0 & k_1 \\ 0 & 0 & k_2 & 0 \\ 0 & k_3 & 0 & 0 \\ k_4 & 0 & 0 & 0 \end{bmatrix}$
(c) $\begin{bmatrix} k & 0 & 0 & 0 \\ 1 & k & 0 & 0 \\ 0 & 1 & k & 0 \\ 0 & 0 & 1 & k \end{bmatrix}$

1.7 FURTHER RESULTS ON SYSTEMS OF EQUATIONS AND INVERTIBILITY

In this section we shall establish more results about systems of linear equations and invertibility of matrices. The first theorem shows how the inverse of the coefficient matrix can be used to solve a system of n equations in n unknowns.

Theorem 1. If A is an invertible $n \times n$ matrix, then for each $n \times 1$ matrix B, the system of equations $AX = B$ has exactly one solution, namely, $X = A^{-1}B$.

Example 1

Consider the system of linear equations

$$x_1 + 2x_2 + 3x_3 = 5$$
$$2x_1 + 5x_2 + 3x_3 = 3$$
$$x_1 \qquad\quad + 8x_3 = 17$$

In matrix form this system can be written as AX = B, where

$$A = \begin{bmatrix} 1 & 2 & 3 \\ 2 & 5 & 3 \\ 1 & 0 & 8 \end{bmatrix} \qquad X = \begin{bmatrix} x_1 \\ x_2 \\ x_3 \end{bmatrix} \qquad B = \begin{bmatrix} 5 \\ 3 \\ 17 \end{bmatrix}$$

In Example 1 of Section 1.6 we showed that A is invertible and

$$A^{-1} = \begin{bmatrix} -40 & 16 & 9 \\ 13 & -5 & -3 \\ 5 & -2 & -1 \end{bmatrix}$$

By Theorem 1 the solution of the system is

$$X = A^{-1}B = \begin{bmatrix} -40 & 16 & 9 \\ 13 & -5 & -3 \\ 5 & -2 & -1 \end{bmatrix} \begin{bmatrix} 5 \\ 3 \\ 17 \end{bmatrix} = \begin{bmatrix} 1 \\ -1 \\ 2 \end{bmatrix}$$

or $x_1 = 1$, $x_2 = -1$, $x_3 = 2$.

The technique illustrated in this example only applies when the coefficient matrix A is square, that is, when the system has as many equations as unknowns. However, many problems in science and engineering involve systems of this type. The method is particularly useful when it is necessary to solve a series of systems

$$AX = B_1, \quad AX = B_2, \quad \ldots, \quad AX = B_k$$

each of which has the same square coefficient matrix A. In this case the solutions

$$X = A^{-1}B_1, \quad X = A^{-1}B_2, \quad \ldots, \quad X = A^{-1}B_k$$

can be obtained using one matrix inversion and k matrix multiplications. This procedure is more efficient than separately applying Gaussian elimination to each of the k systems.

We digress for a moment to illustrate how this situation can arise in applications. In certain applied problems, physical systems are considered that can be described as *black boxes*. This term indicates that the system has been stripped to its bare essentials. As illustrated in Figure 1, one imagines simply that if a certain input is applied to the system, then a certain output for the system will result. The internal workings of the system are either unknown or unimportant for the problem -- hence, the term *black box*. For many important black-box systems, both the input and the output can be described mathematically as matrices having a single column. For example, if the black box consists of certain electronic circuitry, then the input might be an $n \times 1$ matrix whose entries are n voltages read across certain input terminals, and the output might be an $n \times 1$ matrix whose entries are the resulting currents in n output wires. Mathematically speaking, such a system does nothing more than transform an $n \times 1$ input matrix into an $n \times 1$ output matrix. For a large class of black-box systems an input matrix C is related to the output matrix B by a matrix equation

$$AC = B$$

where A is an $n \times n$ matrix whose entries are physical parameters determined by the system. A system of this kind is an example of what is called a *linear physical system*. In applications it is often important to determine what input must be applied to the system to achieve a certain desired output. For a linear physical system of the type we have just discussed, this amounts to solving the equation AX = B for the unknown input X, given the desired output B. Thus if we have a succession of different output matrices $B_1, \ldots, B_k$, and we want to determine the input matrices that produce these given outputs, we must successively solve the k systems of linear equations

$$AX = B_j \qquad j = 1, 2, \ldots, k$$

each of which has the same square coefficient matrix A.

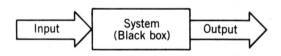

Figure 1

Up to now, to show that an $n \times n$ matrix A is invertible, it has been necessary to find an $n \times n$ matrix B such that

$$AB = I \qquad \text{and} \qquad BA = I$$

The next theorem shows that if we produce an $n \times n$ matrix B satisfying either condition, then the other condition holds automatically.

Theorem 2.   Let A be a square matrix.

(a) If B is a square matrix satisfying BA = I, then B = A^{-1}.

(b) If B is a square matrix satisfying AB = I, then B = A^{-1}.

We are now in a position to add a fourth statement equivalent to the three given in Theorem 1 of Section 1.6.

Theorem 3.   If A is an n × n matrix, then the following statements are equivalent.

(a) A is invertible.

(b) AX = 0 has only the trivial solution.

(c) A is row equivalent to I_n.

(d) AX = B is consistent for every n × 1 matrix B.

EXERCISE SET 1.7

In Exercises 1 - 6 solve the system using the method of Example 1.

1. $x_1 + 2x_2 = 7$

   $2x_1 + 5x_2 = -3$

2. $3x_1 - 6x_2 = 8$

   $2x_1 + 5x_2 = 1$

3. $x_1 + 2x_2 + 2x_3 = -1$

   $x_1 + 3x_2 + x_3 = 4$

   $x_1 + 3x_2 + 2x_3 = 3$

4. $2x_1 + x_2 + x_3 = 7$

   $3x_1 + 2x_2 + x_3 = -3$

   $x_2 + x_3 = 5$

5. $\frac{1}{5}x + \frac{1}{5}y + \frac{1}{5}z = 1$

   $\frac{1}{5}x + \frac{1}{5}y - \frac{4}{5}z = 2$

   $-\frac{2}{5}x + \frac{1}{10}y + \frac{1}{10}z = 0$

6. $3w + x + 7y + 9z = 4$

    $w + x + 4y + 4z = 7$

   $-w \quad - 2y - 3z = 0$

   $-2w - x - 4y - 6z = 6$

7. Solve the system

$$x_1 + 2x_2 + x_3 = b_1$$
$$x_1 - x_2 + x_3 = b_2$$
$$x_1 + x_2 \quad = b_3$$

when

(a) $b_1 = -1$, $b_2 = 3$, $b_3 = 4$

(b) $b_1 = 5$, $b_2 = 0$, $b_3 = 0$

(c) $b_1 = -1$, $b_2 = -1$, $b_3 = 3$

(d) $b_1 = \frac{1}{2}$, $b_2 = 3$, $b_3 = \frac{1}{7}$

8. What conditions must the b's satisfy in order for the given system to be consistent?

(a)
$$x_1 - x_2 + 3x_3 = b_1$$
$$3x_1 - 3x_2 + 9x_3 = b_2$$
$$-2x_1 + 2x_2 - 6x_3 = b_3$$

(b)
$$2x_1 + 3x_2 - x_3 + x_4 = b_1$$
$$x_1 + 5x_2 + x_3 - 2x_4 = b_2$$
$$-x_1 + 2x_2 + 2x_3 - 3x_4 = b_3$$
$$3x_1 + x_2 - 3x_3 + 4x_4 = b_4$$

9. Consider the matrices

$$A = \begin{bmatrix} 2 & 2 & 3 \\ 1 & 2 & 1 \\ 2 & -2 & 1 \end{bmatrix} \quad \text{and} \quad X = \begin{bmatrix} x_1 \\ x_2 \\ x_3 \end{bmatrix}$$

(a) Show that the equation $AX = X$ can be rewritten as $(A - I)X = 0$ and use this result to solve $AX = X$ for X.

(b) Solve $AX = 4X$.

10. Without using pencil and paper, determine if the following matrices are invertible.

(a) $\begin{bmatrix} 2 & 1 & -3 & 1 \\ 0 & 5 & 4 & 3 \\ 0 & 0 & 1 & 2 \\ 0 & 0 & 0 & 3 \end{bmatrix}$ (b) $\begin{bmatrix} 5 & 1 & 4 & 1 \\ 0 & 0 & 2 & -1 \\ 0 & 0 & 1 & 1 \\ 0 & 0 & 0 & 7 \end{bmatrix}$

Hint. Consider the associated homogeneous systems

$$2x_1 + x_2 - 3x_3 + x_4 = 0$$
$$5x_2 + 4x_3 + 3x_4 = 0$$
$$x_3 + 2x_4 = 0$$
$$3x_4 = 0$$

and

$$5x_1 + x_2 + 4x_3 + x_4 = 0$$
$$2x_3 - x_4 = 0$$
$$x_3 + x_4 = 0$$
$$7x_4 = 0$$

11. Let $AX = 0$ be a homogeneous system of n linear equations in n unknowns that has only the trivial solution. Show that if k is any positive integer, then the system $A^k X = 0$ also has only the trivial solution.

12. Let $AX = 0$ be a homogeneous system of n linear equations in n unknowns, and let Q be an invertible matrix. Show that $AX = 0$ has just the trivial solution if and only if $(QA)X = 0$ has just the trivial solution.

SUPPLEMENTARY EXERCISES

1. Use Gauss-Jordan elimination to solve for x' and y' in terms of x and y.

$$x = \frac{3}{5} x' - \frac{4}{5} y'$$
$$y = \frac{4}{5} x' + \frac{3}{5} y'$$

2. Use Gauss-Jordan elimination to solve for x' and y' in terms of x and y.

$$x = x' \cos \theta - y' \sin \theta$$
$$y = x' \sin \theta + y' \cos \theta$$

3. A box containing pennies, nickels, and dimes has 13 coins with a total value of 83 cents. How many coins of each type are in the box?

4. Find positive integers that satisfy

$$x + y + z = 9$$
$$x + 5y + 10z = 44$$

5. For which value(s) of a does the following system have zero, one, infinitely many solutions?

$$x_1 + x_2 + x_3 = 4$$
$$x_3 = 2$$
$$(a^2 - 4)x_3 = a - 2$$

6. Let

$$\begin{bmatrix} a & 0 & b & 2 \\ a & a & 4 & 4 \\ 0 & a & 2 & b \end{bmatrix}$$

be the augmented matrix for a linear system. For what values of a and b does the system have

(a) a unique solution;   (b) a one-parameter solution;

(c) a two-parameter solution;   (d) no solution?

7. Find a matrix K such that AKB = C given that

$$A = \begin{bmatrix} 1 & 4 \\ -2 & 3 \\ 1 & -2 \end{bmatrix}, \quad B = \begin{bmatrix} 2 & 0 & 0 \\ 0 & 1 & -1 \end{bmatrix}, \quad C = \begin{bmatrix} 8 & 6 & -6 \\ 6 & -1 & 1 \\ -4 & 0 & 0 \end{bmatrix}$$

8. If A is $m \times n$ and B is $n \times p$, how many multiplication operations and how many addition operations are needed to calculate the matrix product AB?

9. Let A be a square matrix.

   (a) Show that $(I - A)^{-1} = I + A + A^2 + A^3$ if $A^4 = 0$.

   (b) Show that $(I - A)^{-1} = I + A + A^2 + \cdots + A^n$ if $A^{n+1} = 0$.

10. Find values of a, b, and c so that the graph of the polynomial $p(x) = ax^2 + bx + c$ passes through the points $(1,2)$, $(-1,6)$, and $(2,3)$.

11. Find values of a, b, and c so that the graph of the polynomial $p(x) = ax^2 + bx + c$ passes through the point $(-1,0)$ and has a horizontal tangent at $(2, -9)$.

12. Prove: If A is an $m \times n$ matrix and B is the $n \times 1$ matrix each of whose entries is $1/n$, then

$$AB = \begin{bmatrix} \bar{r}_1 \\ \bar{r}_2 \\ \vdots \\ \bar{r}_m \end{bmatrix}$$

   where $\bar{r}_i$ is the average of the entries in the ith row of A.

13. If

$$C = \begin{bmatrix} c_{11}(x) & c_{12}(x) & \cdots & c_{1n}(x) \\ c_{21}(x) & c_{22}(x) & \cdots & c_{2n}(x) \\ \vdots & \vdots & & \vdots \\ c_{m1}(x) & c_{m2}(x) & \cdots & c_{mn}(x) \end{bmatrix}$$

   where $c_{ij}(x)$ is a differentiable function of x, then we define

$$\frac{dC}{dx} = \begin{bmatrix} c'_{11}(x) & c'_{12}(x) & \cdots & c'_{1n}(x) \\ c'_{21}(x) & c'_{22}(x) & \cdots & c'_{2n}(x) \\ \vdots & \vdots & & \vdots \\ c'_{m1}(x) & c'_{m2}(x) & \cdots & c'_{mn}(x) \end{bmatrix}$$

Show that if the entries in A and B are differentiable functions of x and the product AB is defined, then

$$\frac{d}{dx}(AB) = \frac{dA}{dx}B + A\frac{dB}{dx}$$

14. Find the values of A, B, and C that will make the equation

$$\frac{x^2 + x - 2}{(3x - 1)(x^2 + 1)} = \frac{A}{3x - 1} + \frac{Bx + C}{x^2 + 1}$$

into an identity. [*Hint:* Multiply through by $(3x - 1)(x^2 + 1)$ and equate the corresponding coefficients of the polynomials on each side of the resulting equation.]

2 Determinants

Some discussion of determinants appeared in Section 15.4 of the text. However, all the relevant definitions will be repeated here, so that that section of the text is <u>not</u> a prerequisite for this chapter.

2.1 DEFINITIONS

In this section we will associate with each square matrix A a number called the *determinant* of A. Determinants have important applications to systems of linear equations and can be used to produce a formula for the inverse of an invertible matrix.

The *determinant* of a square matrix A is denoted by det(A) or $|A|$. If A is a 1 × 1 matrix

$$A = \begin{bmatrix} a_{11} \end{bmatrix}$$

then we define

$$\det(A) = a_{11}$$

Example 1

If

$$A = [-3]$$

then

$$\det(A) = \det[-3] = -3$$

If A is a 2×2 matrix

$$A = \begin{bmatrix} a_{11} & a_{12} \\ a_{21} & a_{22} \end{bmatrix}$$

then we define

$$\det(A) = \begin{vmatrix} a_{11} & a_{12} \\ a_{21} & a_{22} \end{vmatrix} = a_{11}a_{22} - a_{12}a_{21} \tag{1}$$

This formula can be remembered with the help of Figure 1. The formula results by multiplying the entries on the rightward arrow and subtracting the product of the entries on the leftward arrow.

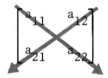

Figure 1

Example 2

If

$$A = \begin{bmatrix} 1 & 5 \\ -2 & 3 \end{bmatrix}$$

then

$$\det(A) = \begin{vmatrix} 1 & 5 \\ -2 & 3 \end{vmatrix} = (1)(3) - (5)(-2) = 3 + 10 = 13$$

If A is a 3×3 matrix

$$A = \begin{bmatrix} a_{11} & a_{12} & a_{13} \\ a_{21} & a_{22} & a_{23} \\ a_{31} & a_{32} & a_{33} \end{bmatrix} \tag{2}$$

then we define

$$\det(A) = \begin{vmatrix} a_{11} & a_{12} & a_{13} \\ a_{21} & a_{22} & a_{23} \\ a_{31} & a_{32} & a_{33} \end{vmatrix} = a_{11} \begin{vmatrix} a_{22} & a_{23} \\ a_{32} & a_{33} \end{vmatrix} - a_{12} \begin{vmatrix} a_{21} & a_{23} \\ a_{31} & a_{33} \end{vmatrix} + a_{13} \begin{vmatrix} a_{21} & a_{22} \\ a_{31} & a_{32} \end{vmatrix} \tag{3}$$

Example 3

$$\begin{vmatrix} 2 & 4 & -3 \\ 1 & 0 & 4 \\ 2 & -1 & 2 \end{vmatrix} = 2 \begin{vmatrix} 0 & 4 \\ -1 & 2 \end{vmatrix} - 4 \begin{vmatrix} 1 & 4 \\ 2 & 2 \end{vmatrix} + (-3) \begin{vmatrix} 1 & 0 \\ 2 & -1 \end{vmatrix}$$

$$= 2(4) - 4(-6) + (-3)(-1) = 35$$

There is a useful alternate procedure for evaluating 3×3 determinants that we will call the "duplicate column method". To see the basis for this method, let us evaluate the 2×2 determinants in (3). This yields

$$\begin{vmatrix} a_{11} & a_{12} & a_{13} \\ a_{21} & a_{22} & a_{23} \\ a_{31} & a_{32} & a_{33} \end{vmatrix} = a_{11}a_{22}a_{33} + a_{12}a_{23}a_{31} + a_{13}a_{21}a_{32} \tag{4}$$
$$- a_{13}a_{22}a_{31} - a_{12}a_{21}a_{33} - a_{11}a_{23}a_{32}$$

The duplicate-column method uses the fact that this formula can be obtained by recopying the first two columns to the right of the determinant sign (Figure 2) then summing the products of the entries on the rightward arrows, and subtracting the products of the entries along the leftward arrows (verify).

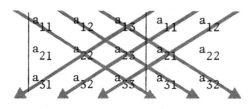

Figure 2

Example 4

Applying the duplicate column method to the determinant in Example 3 yields

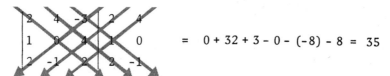 $= 0 + 32 + 3 - 0 - (-8) - 8 = 35$

which agrees with our previous result.

In order to define determinants of higher-order square matrices we will need some preliminary results.

Definition. If A is a square matrix, then the *minor of entry* a_{ij} is denoted by M_{ij} and is defined to be the determinant of the submatrix that remains after the ith row and jth column are deleted from A. The number $(-1)^{i+j}M_{ij}$ is denoted by C_{ij} and is called the *cofactor of entry* a_{ij}.

Example 5

Let

$$A = \begin{bmatrix} 3 & 1 & -4 \\ 2 & 5 & 6 \\ 1 & 4 & 8 \end{bmatrix}$$

The minor of entry a_{11} is

$$M_{11} = \begin{vmatrix} 3 & 1 & -4 \\ 2 & 5 & 6 \\ 1 & 4 & 8 \end{vmatrix} = \begin{vmatrix} 5 & 6 \\ 4 & 8 \end{vmatrix} = 16$$

The cofactor of a_{11} is

$$C_{11} = (-1)^{1+1}M_{11} = M_{11} = 16$$

Similarly, the minor of entry a_{32} is

$$M_{32} = \begin{vmatrix} 3 & 1 & -4 \\ 2 & 5 & 6 \\ 1 & 4 & 8 \end{vmatrix} = \begin{vmatrix} 3 & -4 \\ 2 & 6 \end{vmatrix} = 26$$

The cofactor of a_{32} is

$$C_{32} = (-1)^{3+2}M_{32} = -M_{32} = -26$$

Notice that the cofactor and the minor of an element a_{ij} differ only in sign, that is, $C_{ij} = \pm M_{ij}$. A quick way for determining whether to use the + or − is to use the fact that the sign relating C_{ij} and M_{ij} is in the ith row and jth column of the array

$$\begin{bmatrix} + & - & + & - & + & \cdots \\ - & + & - & + & - & \cdots \\ + & - & + & - & + & \cdots \\ - & + & - & + & - & \cdots \\ \vdots & \vdots & \vdots & \vdots & \vdots & \end{bmatrix}$$

For example, $C_{11} = M_{11}$, $C_{21} = -M_{21}$, $C_{12} = -M_{12}$, $C_{22} = M_{22}$, etc.

For the matrix

$$\begin{bmatrix} a_{11} & a_{12} \\ a_{21} & a_{22} \end{bmatrix}$$

the cofactors of entries a_{11} and a_{12} are

$$C_{11} = \det[a_{22}] = a_{22} \qquad C_{12} = -\det[a_{21}] = -a_{21}$$

so Formula (1) can be written

$$\begin{vmatrix} a_{11} & a_{12} \\ a_{21} & a_{22} \end{vmatrix} = a_{11}C_{11} + a_{12}C_{12} \qquad (6)$$

which states that the determinant of a 2×2 matrix can be computed by multiplying the entries in the first row of A by their cofactors and adding the resulting products. Similarly, for the matrix

$$\begin{bmatrix} a_{11} & a_{12} & a_{13} \\ a_{21} & a_{22} & a_{23} \\ a_{31} & a_{32} & a_{33} \end{bmatrix}$$

the cofactors of entries a_{11}, a_{12}, and a_{13} are

$$C_{11} = \begin{vmatrix} a_{22} & a_{23} \\ a_{32} & a_{33} \end{vmatrix}, \qquad C_{12} = - \begin{vmatrix} a_{21} & a_{23} \\ a_{31} & a_{33} \end{vmatrix}, \qquad C_{13} = \begin{vmatrix} a_{21} & a_{22} \\ a_{31} & a_{32} \end{vmatrix}$$

Thus, Formula (4) can be written

$$\begin{vmatrix} a_{11} & a_{12} & a_{13} \\ a_{21} & a_{22} & a_{23} \\ a_{31} & a_{32} & a_{33} \end{vmatrix} = a_{11}C_{11} + a_{12}C_{12} + a_{13}C_{13} \qquad (7)$$

so just as in the 2×2 case, the determinant of a 3×3 matrix can be obtained by multiplying the entries in the first row by their cofactors and adding the resulting products.

Motivated by Formulas (6) and (7) we define the determinant of an arbitrary $n \times n$ matrix ($n > 1$) as follows:

Definition. If A is an $n \times n$ matrix ($n > 1$) then the *determinant* of A, denoted by $|A|$ or det(A), is defined by

$$\det(A) = a_{11}C_{11} + a_{12}C_{12} + \cdots + a_{1n}C_{1n}$$

where C_{11}, C_{12}, $\ldots$, C_{1n} are the cofactors of the entries in the first row of A.

Example 6

Evaluate det(A), where

$$A = \begin{bmatrix} 1 & 0 & 2 & -3 \\ 3 & 4 & 0 & 1 \\ -1 & 5 & 2 & -2 \\ 0 & 1 & 1 & 3 \end{bmatrix}$$

Solution

$$\det(A) = (1) \begin{vmatrix} 4 & 0 & 1 \\ 5 & 2 & -2 \\ 1 & 1 & 3 \end{vmatrix} - (0) \begin{vmatrix} 3 & 0 & 1 \\ -1 & 2 & -2 \\ 0 & 1 & 3 \end{vmatrix} + (2) \begin{vmatrix} 3 & 4 & 1 \\ -1 & 5 & -2 \\ 0 & 1 & 3 \end{vmatrix}$$

$$- (-3) \begin{vmatrix} 3 & 4 & 0 \\ -1 & 5 & 2 \\ 0 & 1 & 1 \end{vmatrix}$$

$$= (1)(35) - 0 + (2)(62) - (-3)(13) = 198$$

By rearranging the terms in (4) in various ways, it is possible to obtain other formulas like (7). There should be no trouble checking that all of the following are correct (see Exercise 15):

$$\det(A) = a_{11}C_{11} + a_{12}C_{12} + a_{13}C_{13}$$
$$= a_{11}C_{11} + a_{21}C_{21} + a_{31}C_{31}$$
$$= a_{21}C_{21} + a_{22}C_{22} + a_{23}C_{23}$$
$$= a_{12}C_{12} + a_{22}C_{22} + a_{32}C_{32} \qquad (8)$$
$$= a_{31}C_{31} + a_{32}C_{32} + a_{33}C_{33}$$
$$= a_{13}C_{13} + a_{23}C_{23} + a_{33}C_{33}$$

Notice that in each equation the entries and cofactors all come from the same row or column. These equations are called the *cofactor expansions* of det(A).

The results we have just given for 3×3 matrices form a special case of the following general theorem, which we state without proof.

Theorem 1. The determinant of an $n \times n$ matrix A can be computed by multiplying the entries in any row (or column) by their cofactors and adding the resulting products; that is, for each $1 \le i \le n$ and $1 \le j \le n$,

$$\det(A) = a_{1j}C_{1j} + a_{2j}C_{2j} + \cdots + a_{nj}C_{nj}$$

(cofactor expansion along the jth column)

and

$$\det(A) = a_{i1}C_{i1} + a_{i2}C_{i2} + \cdots + a_{in}C_{in}$$

(cofactor expansion along the ith row)

Example 7

Evaluate

$$\begin{vmatrix} 2 & 4 & -3 \\ 1 & 0 & 4 \\ 2 & -1 & 2 \end{vmatrix}$$

by a cofactor expansion along the second column.

Solution

$$\begin{vmatrix} 2 & 4 & -3 \\ 1 & 0 & 4 \\ 2 & -1 & 2 \end{vmatrix} = -(4) \begin{vmatrix} 1 & 4 \\ 2 & 2 \end{vmatrix} + 0 \begin{vmatrix} 2 & -3 \\ 2 & 2 \end{vmatrix} - (-1) \begin{vmatrix} 2 & -3 \\ 1 & 4 \end{vmatrix}$$

$$= -4(-6) + 0 + (1)(11) = 35$$

This agrees with the results in Examples 3 and 4.

REMARK. In this example it was unnecessary to compute the second cofactor, since it was multiplied by zero. In general, the best strategy for evaluating a determinant by cofactor expansion is to expand along a row or column having the largest number of zeros.

The following result is a direct consequence of Theorem 1.

Theorem 2. If A is a square matrix with a row or column of zeros, then det(A) = 0.

Proof. Evaluate det(A) by a cofactor expansion along the row or column of zeros.

EXERCISE SET 2.1

1. Find det(A) if

   (a) A = [2]   (b) A = [-7]

2. Let

$$A = \begin{bmatrix} -1 & 2 & 3 \\ 4 & 1 & -6 \\ -3 & 5 & 2 \end{bmatrix}$$

   Find det(A) using

   (a) the duplicate column method

   (b) cofactor expansion along the first row

   (c) cofactor expansion along the third row

(d) cofactor expansion along the first column

(e) cofactor expansion along the second column.

In Exercises 3 - 14 evaluate the determinant.

3. $\begin{vmatrix} 1 & 2 \\ -1 & 3 \end{vmatrix}$ 4. $\begin{vmatrix} 6 & 4 \\ 3 & 2 \end{vmatrix}$ 5. $\begin{vmatrix} -1 & 7 \\ -8 & -3 \end{vmatrix}$ 6. $\begin{vmatrix} k-1 & 2 \\ 4 & k-3 \end{vmatrix}$

7. $\begin{vmatrix} 1 & -2 & 7 \\ 3 & 5 & 1 \\ 4 & 3 & 8 \end{vmatrix}$ 8. $\begin{vmatrix} 8 & 2 & -1 \\ -3 & 4 & -6 \\ 1 & 7 & 2 \end{vmatrix}$ 9. $\begin{vmatrix} 1 & 0 & 3 \\ 4 & 0 & -1 \\ 2 & 8 & 6 \end{vmatrix}$ 10. $\begin{vmatrix} k & -3 & 9 \\ 2 & 4 & k+1 \\ 1 & k^2 & 3 \end{vmatrix}$

11. $\begin{vmatrix} 1 & 4 & -3 & 1 \\ 2 & 0 & 6 & 3 \\ 4 & -1 & 2 & 5 \\ 1 & 0 & -2 & 4 \end{vmatrix}$ 12. $\begin{vmatrix} 2 & -1 & 4 & 1 \\ 3 & 5 & 3 & -2 \\ 4 & -2 & 0 & 2 \\ 3 & 3 & -1 & 5 \end{vmatrix}$

13. $\begin{vmatrix} 0 & 0 & 0 & 0 & 1 \\ 0 & 0 & 0 & 2 & 0 \\ 0 & 0 & 3 & 0 & 0 \\ 0 & 4 & 0 & 0 & 0 \\ 5 & 0 & 0 & 0 & 0 \end{vmatrix}$ 14. $\begin{vmatrix} 0 & 4 & 0 & 0 & 0 \\ 0 & 0 & 0 & 0 & 0 \\ 0 & 0 & 3 & 0 & 0 \\ 0 & 0 & 0 & 0 & 1 \\ 5 & 0 & 0 & 0 & 0 \end{vmatrix}$

15. Derive the first and last cofactor expansions listed in (8).

2.2 PROPERTIES OF DETERMINANTS; ALTERNATE METHODS OF EVALUATION

Cofactor expansion is not a workable method for evaluating determinants of large matrices. For example, the evaluation of a 10×10 determinant by cofactor expansion requires 3,628,799 additions and 32,659,200 multiplications. Even the fastest of our current digital computers cannot handle the computation of a 25×25 determinant by this method in a practical amount of time. This section is devoted, therefore, to developing properties of determinants that will simplify their evaluation.

We begin with a basic theorem that shows how an elementary row operation on a matrix affects the value of its determinant.

Theorem 1. Let A be any $n \times n$ matrix.

(a) If A' is the matrix that results when a single row of A is multiplied by a constant k, then $\det(A') = k \det(A)$.

(b) If A' is the matrix that results when two rows of A are interchanged, then $\det(A') = -\det(A)$.

(c) If A' is the matrix that results when a multiple of one row of A is added to another row, then $\det(A') = \det(A)$.

The following examples illustrate this theorem for 2×2 matrices.

Example 1

Multiplying the first row of

$$A = \begin{bmatrix} a_{11} & a_{12} \\ a_{21} & a_{22} \end{bmatrix}$$

by k yields

$$A' = \begin{bmatrix} ka_{11} & ka_{12} \\ a_{21} & a_{22} \end{bmatrix}$$

so

$$\det(A') = (ka_{11})a_{22} - (ka_{12})a_{22} = k(a_{11}a_{22} - a_{12}a_{21}) = k \det(A)$$

Example 2

Interchanging the rows of

$$A = \begin{bmatrix} a_{11} & a_{12} \\ a_{21} & a_{22} \end{bmatrix}$$

yields

$$A' = \begin{bmatrix} a_{21} & a_{22} \\ a_{11} & a_{12} \end{bmatrix}$$

so

$$\det(A') = a_{21}a_{12} - a_{22}a_{11} = -(a_{11}a_{22} - a_{12}a_{21}) = -\det(A)$$

Example 3

Adding the k times the first row of

$$A = \begin{bmatrix} a_{11} & a_{12} \\ a_{21} & a_{22} \end{bmatrix}$$

to the second row yields

$$A' = \begin{bmatrix} a_{11} & a_{12} \\ a_{21} + ka_{11} & a_{22} + ka_{12} \end{bmatrix}$$

so

$$\det(A') = a_{11}(a_{22} + ka_{12}) - a_{12}(a_{21} + ka_{11})$$
$$= (a_{11}a_{22} - a_{12}a_{21}) + k(a_{11}a_{12} - a_{11}a_{12})$$
$$= a_{11}a_{22} - a_{12}a_{21} = \det(A)$$

Statement (a) in Theorem 1 has an alternate interpretation that is sometimes useful. This result allows us to bring a "common factor" from any row of a square matrix through the determinant sign. To illustrate, consider the matrices

$$A = \begin{bmatrix} a_{11} & a_{12} & a_{13} \\ a_{21} & a_{22} & a_{23} \\ a_{31} & a_{32} & a_{33} \end{bmatrix} \qquad B = \begin{bmatrix} a_{11} & a_{12} & a_{13} \\ ka_{21} & ka_{22} & ka_{23} \\ a_{31} & a_{32} & a_{33} \end{bmatrix}$$

where the second row of B has a common factor of k. Since B is the matrix that results when the second row of A is multiplied by k, statement (a) in Theorem 1 asserts that $\det(B) = k \det(A)$; that is,

$$\begin{vmatrix} a_{11} & a_{12} & a_{13} \\ ka_{21} & ka_{22} & ka_{23} \\ a_{31} & a_{32} & a_{33} \end{vmatrix} = k \begin{vmatrix} a_{11} & a_{12} & a_{13} \\ a_{21} & a_{22} & a_{23} \\ a_{31} & a_{32} & a_{33} \end{vmatrix}$$

Theorem 2. If A is a square matrix and all the entries above or below the main diagonal are zero, then det(A) is the product of the entries on the main diagonal.

Example 4

$$\begin{vmatrix} 2 & 7 & -3 & 8 & 3 \\ 0 & -3 & 7 & 5 & 1 \\ 0 & 0 & 6 & 7 & 6 \\ 0 & 0 & 0 & 9 & 8 \\ 0 & 0 & 0 & 0 & 4 \end{vmatrix} = (2)(-3)(6)(9)(4) = -1296$$

Example 5

Since I_n (the $n \times n$ identity matrix) has 1's along the main diagonal and zeros above and below the main diagonal, it follows that for all n

$$\det(I_n) = 1$$

We shall now formulate an alternate method for evaluating determinants that will avoid the large amount of computation involved in cofactor expansion. The basic idea of this method is to apply elementary row operations to reduce the given matrix A to a matrix R that is in row-echelon form. Since R will have all zeros below the main diagonal, det(R) can be obtained from Theorem 2 and det(A) can then be obtained by using Theorem 1 to relate the unknown value of det(A) to the known value of det(R). The following example illustrates this method.

Example 5

Evaluate det(A) where

$$A = \begin{bmatrix} 0 & 1 & 5 \\ 3 & -6 & 9 \\ 2 & 6 & 1 \end{bmatrix}$$

Solution. Reducing A to row-echelon form and applying Theorem 1, we obtain

$$\det(A) = \begin{vmatrix} 0 & 1 & 5 \\ 3 & -6 & 9 \\ 2 & 6 & 1 \end{vmatrix} = - \begin{vmatrix} 3 & -6 & 9 \\ 0 & 1 & 5 \\ 2 & 6 & 1 \end{vmatrix}$$

> The first and second rows of A were interchanged.

$$= -3 \begin{vmatrix} 1 & -2 & 3 \\ 0 & 1 & 5 \\ 2 & 6 & 1 \end{vmatrix}$$

> A common factor of 3 from the first row of the preceding matrix was taken through the det sign.

$$= -3 \begin{vmatrix} 1 & -2 & 3 \\ 0 & 1 & 5 \\ 0 & 10 & -5 \end{vmatrix}$$

> -2 times the first row of the preceding matrix was added to the third row.

$$= -3 \begin{vmatrix} 1 & -2 & 3 \\ 0 & 1 & 5 \\ 0 & 0 & -55 \end{vmatrix}$$

> -10 times the second row of the preceding matrix was added to the third row.

$$= (-3)(-55) \begin{vmatrix} 1 & -2 & 3 \\ 0 & 1 & 5 \\ 0 & 0 & 1 \end{vmatrix}$$

> A common factor of -55 from the last row of the preceding matrix was taken through the det sign.

$$= (-3)(-55)(1) = 165$$

REMARK. The method of row reduction is well suited for computer evaluation of determinants because it is systematic and easily programmed.

Example 6

Evaluate det(A), where

$$A = \begin{bmatrix} 1 & 3 & -2 & 4 \\ 2 & 6 & -4 & 8 \\ 3 & 9 & 1 & 5 \\ 1 & 1 & 4 & 8 \end{bmatrix}$$

$$
\det(A) = \begin{vmatrix} 1 & 3 & -2 & 4 \\ 0 & 0 & 0 & 0 \\ 3 & 9 & 1 & 5 \\ 1 & 1 & 4 & 8 \end{vmatrix}
$$

-2 times the first row of A was added to the second row.

No further reduction is needed since it follows from Theorem 2 of Section 2.1 that det(A) = 0.

It should be evident from this example that whenever a square matrix has two proportional rows (like the first and second rows of A), it is possible to introduce a row of zeros by adding a suitable multiple of one of these rows to the other. Thus, *if a square matrix has two proportional rows, its determinant is zero.*

Example 7

Each of the following matrices has two proportional rows; thus, by inspection, each has a zero determinant.

$$
\begin{bmatrix} -1 & 4 \\ -2 & 8 \end{bmatrix}
\qquad
\begin{bmatrix} 2 & 7 & 8 \\ 3 & 2 & 4 \\ 2 & 7 & 8 \end{bmatrix}
\qquad
\begin{bmatrix} 3 & -1 & 4 & -5 \\ 6 & -2 & 5 & 2 \\ 5 & 8 & 1 & 4 \\ -9 & 3 & -12 & 15 \end{bmatrix}
$$

Cofactor expansion and row or column operations can sometimes be used in combination to provide an effective method for evaluating determinants. The following example illustrates this idea.

Example 8

Evaluate det(A) where

$$
A = \begin{bmatrix} 3 & 5 & -2 & 6 \\ 1 & 2 & -1 & 1 \\ 2 & 4 & 1 & 5 \\ 3 & 7 & 5 & 3 \end{bmatrix}
$$

Solution. By adding suitable multiples of the second row to the remaining rows, we obtain

$$\det(A) = \begin{vmatrix} 0 & -1 & 1 & 3 \\ 1 & 2 & -1 & 1 \\ 0 & 0 & 3 & 3 \\ 0 & 1 & 8 & 0 \end{vmatrix}$$

$$= - \begin{vmatrix} -1 & 1 & 3 \\ 0 & 3 & 3 \\ 1 & 8 & 0 \end{vmatrix} \qquad \boxed{\begin{array}{l}\text{Cofactor expansion along} \\ \text{the first column.}\end{array}}$$

$$= - \begin{vmatrix} -1 & 1 & 3 \\ 0 & 3 & 3 \\ 0 & 9 & 3 \end{vmatrix} \qquad \boxed{\begin{array}{l}\text{We added the first row} \\ \text{to the third row.}\end{array}}$$

$$= -(-1) \begin{vmatrix} 3 & 3 \\ 9 & 3 \end{vmatrix} \qquad \boxed{\begin{array}{l}\text{Cofactor expansion along} \\ \text{the first column.}\end{array}}$$

$$= -18$$

We conclude this section with a theorem that lists three more major properties of determinants.

Theorem 3.

(a) If A is a square matrix, then $\det(A) = \det(A^t)$

(b) If A and B are square matrices of the same size, then $\det(AB) = \det(A)\det(B)$

(c) A square matrix A is invertible if and only if $\det(A) \neq 0$.

Example 9

Consider the matrices

$$A = \begin{bmatrix} 3 & 1 \\ 2 & 1 \end{bmatrix} \qquad B = \begin{bmatrix} -1 & 3 \\ 5 & 8 \end{bmatrix} \qquad AB = \begin{bmatrix} 2 & 17 \\ 3 & 14 \end{bmatrix}$$

We have det(A)det(B) = (1)(-23) = -23. On the other hand, by direct computation det(AB) = -23, so that det(AB) = det(A)det(B).

Example 10

Since the first and third rows of

$$A = \begin{bmatrix} 1 & 2 & 3 \\ 1 & 0 & 1 \\ 2 & 4 & 6 \end{bmatrix}$$

are proportional, det(A) = 0. Thus A is not invertible.

Because of part (a) of Theorem 3, nearly every theorem about determinants that contains the word "row" in its statement is also true when the word "column" is substituted for "row". To prove a column statement one need only transpose the matrix in question to convert the column statement to a row statement, and then apply the corresponding known result for rows.

Example 11

By inspection, the matrix

$$\begin{bmatrix} 1 & -2 & 7 \\ -4 & 8 & 5 \\ 2 & -4 & 3 \end{bmatrix}$$

has a zero determinant since the first and second columns are proportional.

EXERCISE SET 2.2

1. Evaluate the following by inspection.

(a) $\begin{vmatrix} 2 & -40 & 17 \\ 0 & 1 & 11 \\ 0 & 0 & 3 \end{vmatrix}$

(b) $\begin{vmatrix} 1 & 0 & 0 & 0 \\ -9 & -1 & 0 & 0 \\ 12 & 7 & 8 & 0 \\ 4 & 5 & 7 & 2 \end{vmatrix}$

(c) $\begin{vmatrix} 1 & 2 & 3 \\ 3 & 7 & 6 \\ 1 & 2 & 3 \end{vmatrix}$

(d) $\begin{vmatrix} 3 & -1 & 2 \\ 6 & -2 & 4 \\ 1 & 7 & 3 \end{vmatrix}$

In Exercises 2 - 9 evaluate the determinants of the given matrices by reducing the matrix to row-echelon form.

2. $\begin{bmatrix} 2 & 3 & 7 \\ 0 & 0 & -3 \\ 1 & -2 & 7 \end{bmatrix}$

3. $\begin{bmatrix} 2 & 1 & 1 \\ 4 & 2 & 3 \\ 1 & 3 & 0 \end{bmatrix}$

4. $\begin{bmatrix} 1 & -2 & 0 \\ -3 & 5 & 1 \\ 4 & -3 & 2 \end{bmatrix}$

5. $\begin{bmatrix} 2 & -4 & 8 \\ -2 & 7 & -2 \\ 0 & 1 & 5 \end{bmatrix}$

6. $\begin{bmatrix} 3 & 6 & 9 & 3 \\ -1 & 0 & 1 & 0 \\ 1 & 3 & 2 & -1 \\ -1 & -2 & -2 & 1 \end{bmatrix}$

7. $\begin{bmatrix} 2 & 1 & 3 & 1 \\ 1 & 0 & 1 & 1 \\ 0 & 2 & 1 & 0 \\ 0 & 1 & 2 & 3 \end{bmatrix}$

8. $\begin{bmatrix} \frac{1}{2} & \frac{1}{2} & 1 & \frac{1}{2} \\ -\frac{1}{2} & \frac{1}{2} & 0 & \frac{1}{2} \\ \frac{2}{3} & \frac{1}{3} & \frac{1}{3} & 0 \\ \frac{1}{3} & 1 & \frac{1}{3} & 0 \end{bmatrix}$

9. $\begin{bmatrix} 1 & 3 & 1 & 5 & 3 \\ -2 & -7 & 0 & -4 & 2 \\ 0 & 0 & 1 & 0 & 1 \\ 0 & 0 & 2 & 1 & 1 \\ 0 & 0 & 0 & 1 & 1 \end{bmatrix}$

10. Verify that $\det(A^t) = \det(A)$ if

$$A = \begin{bmatrix} 1 & 2 & 7 \\ -1 & 0 & 6 \\ 3 & 2 & 8 \end{bmatrix}$$

11. Verify that $\det(AB) = \det(A)\det(B)$ if

$$A = \begin{bmatrix} 2 & 1 & 0 \\ 3 & 4 & 0 \\ 0 & 0 & 2 \end{bmatrix} \quad \text{and} \quad B = \begin{bmatrix} 1 & -1 & 3 \\ 7 & 1 & 2 \\ 5 & 0 & 1 \end{bmatrix}$$

12. Use Theorem 3(c) to determine which of the following matrices are invertible.

(a) $\begin{bmatrix} 1 & 0 & 0 \\ 3 & 6 & 7 \\ 0 & 8 & -1 \end{bmatrix}$
(b) $\begin{bmatrix} -2 & 1 & -4 \\ 1 & 1 & 2 \\ 3 & 1 & 6 \end{bmatrix}$

(c) $\begin{bmatrix} 7 & 2 & 1 \\ 7 & 2 & 1 \\ 3 & 6 & 6 \end{bmatrix}$
(d) $\begin{bmatrix} 0 & 7 & 5 \\ 0 & 1 & -1 \\ 0 & 3 & 2 \end{bmatrix}$

13. Assume $\det \begin{bmatrix} a & b & c \\ d & e & f \\ g & h & i \end{bmatrix} = 5$. Find

(a) $\det \begin{bmatrix} d & e & f \\ g & h & i \\ a & b & c \end{bmatrix}$
(b) $\det \begin{bmatrix} -a & -b & -c \\ 2d & 2e & 2f \\ -g & -h & -i \end{bmatrix}$

(c) $\det \begin{bmatrix} a+d & b+e & c+f \\ d & e & f \\ g & h & i \end{bmatrix}$
(d) $\det \begin{bmatrix} a & b & c \\ d-3a & e-3b & f-3c \\ 2g & 2h & 2i \end{bmatrix}$

14. Use row reduction to show that

$$\begin{vmatrix} 1 & 1 & 1 \\ a & b & c \\ a^2 & b^2 & c^2 \end{vmatrix} = (b-a)(c-a)(c-b)$$

15. Without directly evaluating, show that $x = 0$ and $x = 2$ satisfy

$$\begin{vmatrix} x^2 & x & 2 \\ 2 & 1 & 1 \\ 0 & 0 & -5 \end{vmatrix} = 0$$

16. For which value(s) of k does A fail to be invertible?

(a) $A = \begin{bmatrix} k-3 & -2 \\ -2 & k-2 \end{bmatrix}$ (b) $A = \begin{bmatrix} 1 & 2 & 4 \\ 3 & 1 & 6 \\ k & 3 & 2 \end{bmatrix}$

17. Let $AX = 0$ be a system of n linear equations in n unknowns. Show that the system has a nontrivial solution if and only if $\det(A) = 0$.

18. Prove the following special cases of Theorem 1.

(a) $\begin{vmatrix} ka_{11} & ka_{12} & ka_{13} \\ a_{21} & a_{22} & a_{23} \\ a_{31} & a_{32} & a_{33} \end{vmatrix} = k \begin{vmatrix} a_{11} & a_{12} & a_{13} \\ a_{21} & a_{22} & a_{23} \\ a_{31} & a_{32} & a_{33} \end{vmatrix}$

(b) $\begin{vmatrix} a_{21} & a_{22} & a_{23} \\ a_{11} & a_{12} & a_{13} \\ a_{31} & a_{32} & a_{33} \end{vmatrix} = - \begin{vmatrix} a_{11} & a_{12} & a_{13} \\ a_{21} & a_{22} & a_{23} \\ a_{31} & a_{32} & a_{33} \end{vmatrix}$

(c) $\begin{vmatrix} a_{11} + ka_{21} & a_{12} + ka_{22} & a_{13} + ka_{23} \\ a_{21} & a_{22} & a_{23} \\ a_{31} & a_{32} & a_{33} \end{vmatrix} = \begin{vmatrix} a_{11} & a_{12} & a_{13} \\ a_{21} & a_{22} & a_{23} \\ a_{31} & a_{32} & a_{33} \end{vmatrix}$

19. Show that if A is an $n \times n$ matrix then $\det(kA) = k^n \det(A)$.

20. Find 2×2 matrices A and B such that $\det(A + B) \neq \det(A) + \det(B)$.

2.3 ADJOINT FORMULA FOR A^{-1}; CRAMER'S RULE

In Chapter 1 we gave algorithms for solving systems of linear equations and for calculating the inverse of a matrix. In this section we shall obtain a formula for the solution of certain systems of linear equations in terms of determinants and a formula for the inverse of an invertible matrix.

Definition. If A is any $n \times n$ matrix and C_{ij} is the cofactor of a_{ij}, then the matrix

$$\begin{bmatrix} C_{11} & C_{12} & \cdots & C_{1n} \\ C_{21} & C_{22} & \cdots & C_{2n} \\ \vdots & \vdots & & \vdots \\ C_{n1} & C_{n2} & \cdots & C_{nn} \end{bmatrix}$$

is called the *matrix of cofactors from A*. The transpose of this matrix is called the *adjoint of* A and is denoted by adj(A).

Example 1

Let

$$A = \begin{bmatrix} 3 & 2 & -1 \\ 1 & 6 & 3 \\ 2 & -4 & 0 \end{bmatrix}$$

The cofactors of A are

$$C_{11} = 12 \qquad C_{12} = 6 \qquad C_{13} = -16$$
$$C_{21} = 4 \qquad C_{22} = 2 \qquad C_{23} = 16$$
$$C_{31} = 12 \qquad C_{32} = -10 \qquad C_{33} = 16$$

so that the matrix of cofactors is

$$\begin{bmatrix} 12 & 6 & -16 \\ 4 & 2 & 16 \\ 12 & -10 & 16 \end{bmatrix}$$

and the adjoint of A is

$$\text{adj}(A) = \begin{bmatrix} 12 & 4 & 12 \\ 6 & 2 & -10 \\ -16 & 16 & 16 \end{bmatrix}$$

We are now in position to state a formula for the inverse of an invertible matrix.

Theorem 1. If A is an invertible matrix, then

$$A^{-1} = \frac{1}{\det(A)} \text{adj}(A) \tag{1}$$

Example 2

Use (1) to find the inverse of the matrix A in Example 1.

Solution. The reader can check that $\det(A) = 64$. Thus

$$A^{-1} = \frac{1}{\det(A)} \text{adj}(A) = \frac{1}{64} \begin{bmatrix} 12 & 4 & 12 \\ 6 & 2 & -10 \\ -16 & 16 & 16 \end{bmatrix}$$

$$= \begin{bmatrix} \frac{12}{64} & \frac{4}{64} & \frac{12}{64} \\ \frac{6}{64} & \frac{2}{64} & -\frac{10}{64} \\ -\frac{16}{64} & \frac{16}{64} & \frac{16}{64} \end{bmatrix}$$

We note that for matrices larger than 3×3 the matrix inversion method in this example is less efficient than the technique given in Section 1.6. On the other hand, the inversion method in Section 1.6 is just a computational procedure or algorithm for computing A^{-1} and is not very useful for studying properties of the inverse. Formula (1) can often be used to obtain properties of the inverse without actually computing it (Exercise 13).

In a similar vein, it is often useful to have a formula for the solution of a system of equations that can be used to study properties of the solution without solving the system. The next theorem establishes such a formula for systems of n equations in n unknowns. The formula is known as *Cramer's rule*.

Theorem 2. *(Cramer's Rule).* If AX = B is a system of n linear equations in n unknowns such that $\det(A) \neq 0$, then the system has a unique solution. This solution is

$$x_1 = \frac{\det(A_1)}{\det(A)}, \quad x_2 = \frac{\det(A_2)}{\det(A)}, \quad \ldots, \quad x_n = \frac{\det(A_n)}{\det(A)}$$

where A_j is the matrix obtained by replacing the entries in the jth column of A by the entries in the matrix

$$B = \begin{bmatrix} b_1 \\ b_2 \\ \vdots \\ b_n \end{bmatrix}$$

Example 3

Use Cramer's Rule to solve

$$x_1 + + 2x_3 = 6$$
$$-3x_1 + 4x_2 + 6x_3 = 30$$
$$-x_1 - 2x_2 + 3x_3 = 8$$

Solution.

$$A = \begin{bmatrix} 1 & 0 & 2 \\ -3 & 4 & 6 \\ -1 & -2 & 3 \end{bmatrix} \qquad A_1 = \begin{bmatrix} 6 & 0 & 2 \\ 30 & 4 & 6 \\ 8 & -2 & 3 \end{bmatrix}$$

$$A_2 = \begin{bmatrix} 1 & 6 & 2 \\ -3 & 30 & 6 \\ -1 & 8 & 3 \end{bmatrix} \qquad A_3 = \begin{bmatrix} 1 & 0 & 6 \\ -3 & 4 & 30 \\ -1 & -2 & 8 \end{bmatrix}$$

Therefore

$$x_1 = \frac{\det(A_1)}{\det(A)} = \frac{-40}{44} = \frac{-10}{11}, \qquad x_2 = \frac{\det(A_2)}{\det(A)} = \frac{72}{44} = \frac{18}{11},$$

$$x_3 = \frac{\det(A_3)}{\det(A)} = \frac{152}{44} = \frac{38}{11}$$

To solve a system of n equations in n unknowns by Cramer's Rule, it is necessary to evaluate $n+1$ determinants of $n \times n$ matrices. For systems with more than three equations, Gaussian elimination is superior computationally since it is only necessary to reduce one n by $n+1$ augmented matrix. Cramer's rule, however, gives a formula for the solution.

EXERCISE SET 2.3

1. Let

$$A = \begin{bmatrix} 1 & 6 & -3 \\ -2 & 7 & 1 \\ 3 & -1 & 4 \end{bmatrix}$$

Find

(a) the matrix of cofactors

(b) adj (A)

(c) A^{-1} using Formula (1).

In Exercises 2 - 3, use Formula (1) to find the inverse of A.

2. $A = \begin{bmatrix} 0 & 1 & 2 \\ 2 & 4 & 3 \\ 3 & 7 & 6 \end{bmatrix}$

3. $A = \begin{bmatrix} 1 & 0 & 1 \\ -1 & 3 & 0 \\ 1 & 0 & 2 \end{bmatrix}$

4. Let

$$A = \begin{bmatrix} 1 & 3 & 1 & 1 \\ 2 & 5 & 2 & 2 \\ 1 & 3 & 8 & 9 \\ 1 & 3 & 2 & 2 \end{bmatrix}$$

(a) Evaluate A^{-1} using Formula (1).

(b) Evaluate A^{-1} using the method of Example 1 in Section 1.6.

(c) Which method involves less computation?

In Exercises 5 - 10 solve by Cramer's rule, where it applies.

5. $3x_1 - 4x_2 = -5$

$2x_1 + x_2 = 4$

6. $4x + 5y = 2$

$11x + y + 2z = 3$

$x + 5y + 2z = 1$

7. $x + y - 2z = 1$

$2x - y + z = 2$

$x - 2y - 4z = -4$

8. $x_1 - 3x_2 + x_3 = 4$

$2x_1 - x_2 = -2$

$4x_1 - 3x_3 = 0$

9. $2x_1 - x_2 + x_3 - 4x_4 = -32$

$7x_1 + 2x_2 + 9x_3 - x_4 = 14$

$3x_1 - x_2 + x_3 + x_4 = 11$

$x_1 + x_2 - 4x_3 - 2x_4 = -4$

10. $2x_1 - x_2 + x_3 = 8$

$4x_1 + 3x_2 + x_3 = 7$

$6x_1 + 2x_2 + 2x_3 = 15$

11. Use Cramer's rule to solve for z without solving for x, y, and w.

$$4x + y + z + w = 6$$
$$3x + 7y - z + w = 1$$
$$7x + 3y - 5z + 8w = -3$$
$$x + y + z + 2w = 3$$

12. Let AX = B be the system in Exercise 11.

   (a) Solve by Cramer's Rule.

   (b) Solve by Gauss-Jordan elimination.

   (c) Which method involves the least amount of computation?

13. Prove that if det(A) = 1 and all the entries in A are integers, then all the entries in A^{-1} are integers.

14. Let AX = B be a system of n linear equations in n unknowns with integer coefficients and integer constants. Prove that if det(A) = 1, then the solution X has integer entries.

SUPPLEMENTARY EXERCISES

1. Use Cramer's Rule to solve for x' and y' in terms of x and y.

$$x = \frac{3}{5} x' - \frac{4}{5} y'$$
$$y = \frac{4}{5} x' + \frac{3}{5} y'$$

2. Use Cramer's Rule to solve for x' and y' in terms of x and y.

$$x = x' \cos \theta - y' \sin \theta$$
$$y = x' \sin \theta + y' \cos \theta$$

3. By examining the determinant of the coefficient matrix, show that the following system has a nontrivial solution if and only if $\alpha = \beta$.

    $$x + y + \alpha z = 0$$
    $$x + y + \beta z = 0$$
    $$\alpha x + \beta y + z = 0$$

4. Show that if a square matrix A satisfies

    $$A^3 + 4A^2 - 2A + 7I = 0$$

    then so does A^t.

5. (a) For the triangle below, use trigonometry to show

    $$b \cos \gamma + c \cos \beta = a$$
    $$c \cos \alpha + a \cos \gamma = b$$
    $$a \cos \beta + b \cos \alpha = c$$

    and then apply Cramer's Rule to show

    $$\cos \alpha = \frac{b^2 + c^2 - a^2}{2bc}$$

    (b) Use Cramer's Rule to obtain similar formulas for $\cos \beta$ and $\cos \gamma$.

    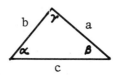

6. Use determinants to show that for all real values of λ the only solution of

    $$x - 2y = \lambda x$$
    $$x - y = \lambda y$$

    is $x = 0$, $y = 0$.

7. Show that if $f_1(x)$, $f_2(x)$, $g_1(x)$, and $g_2(x)$ are differentiable functions, and if

$$W = \begin{vmatrix} f_1(x) & f_2(x) \\ g_1(x) & g_2(x) \end{vmatrix}$$

then

$$\frac{dW}{dx} = \begin{vmatrix} f_1'(x) & f_2'(x) \\ g_1(x) & g_2(x) \end{vmatrix} + \begin{vmatrix} f_1(x) & f_2(x) \\ g_1'(x) & g_2'(x) \end{vmatrix}$$

8. (a) In the figure below, the area of triangle ABC is expressible as

    area ABC = area ADEC + area CEFB - area ADFB

    Use this and the fact that the area of a trapezoid equals $\frac{1}{2}$ the altitude times the sum of the parallel sides to show that

    $$\text{area ABC} = \frac{1}{2} \begin{vmatrix} x_1 & y_1 & 1 \\ x_2 & y_2 & 1 \\ x_3 & y_3 & 1 \end{vmatrix}$$

    [*Note.* In the derivation of this formula, the vertices are labeled so the triangle is traced counterclockwise proceeding from (x_1, y_1) to (x_2, y_2) to (x_3, y_3). For a clockwise orientation, the determinant above yields the *negative* of the area.]

    (b) Use the result in (a) to find the area of the triangle with vertices (3,3), (4,0), (-2,-1).

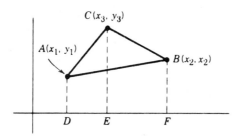

9. Prove: If the entries in each row of an $n \times n$ matrix A add up to zero, then $\det(A) = 0$. [*Hint.* Consider the product AX, where X is the $n \times 1$ matrix, each of whose entries is one.]

10. Prove: The equation of the line through the distinct points (a_1, b_1) and (a_2, b_2) can be written

$$\begin{vmatrix} x & y & 1 \\ a_1 & b_1 & 1 \\ a_2 & b_2 & 1 \end{vmatrix} = 0$$

11. Prove: (x_1, y_1), (x_2, y_2), and (x_3, y_3) are collinear points if and only if

$$\begin{vmatrix} x_1 & y_1 & 1 \\ x_2 & y_2 & 1 \\ x_3 & y_3 & 1 \end{vmatrix} = 0$$

12. Prove: The equation of the plane through the noncollinear points (a_1, b_1, c_1), (a_2, b_2, c_2), and (a_3, b_3, c_3) can be written:

$$\begin{vmatrix} x & y & z & 1 \\ a_1 & b_1 & c_1 & 1 \\ a_2 & b_2 & c_2 & 1 \\ a_3 & b_3 & c_3 & 1 \end{vmatrix} = 0$$

3 Applications to Calculus

3.1 EUCLIDEAN N-DIMENSIONAL SPACE

In analytic geometry a point in 2-space is represented by a pair of real numbers (a, b) and a point in 3-space by a triple of real numbers (a, b, c).

By the latter part of the nineteenth century mathematicians and physicists began to realize that there was no need to stop with triples. It was recognized that quadruples of numbers (a_1, a_2, a_3, a_4) could be regarded as points in "4-dimensional" space, quintuples $(a_1, a_2, \ldots, a_5)$ as points in "5-dimensional" space, etc. Although our geometric visualization does not extend beyond 3-space, it is nevertheless possible to extend many familiar ideas beyond 3-space by working with analytic or numerical properties of points and vectors rather than the geometric properties. In this section we shall make these ideas more precise.

Definition. If   is a positive integer, then an *ordered-n-tuple* is a sequence of n real numbers $(a_1, a_2, \ldots, a_n)$. The set of all ordered n-tuples is called *n-space* and is denoted by R^n.

When n = 2 or 3, it is usual to use the terms *ordered pair* and *ordered triple* rather than ordered 2-tuple and 3-tuple. When n = 1, each ordered n-tuple consists of one real number, and so R^1 may be viewed as the set of real numbers. It is usual to write R rather than R^1 for this set.

It might have occurred to the reader in the study of 3-space that the symbol (a_1, a_2, a_3) has two different geometric interpretations. It can be interpreted as a point, in which case a_1, a_2 and a_3 are the coordinates (Figure 1a) or it can be interpreted as a vector, in which case a_1, a_2 and a_3 are the components (Figure 1b). It follows, therefore, that an ordered n-tuple $(a_1, a_2, \ldots, a_n)$ can be viewed either as a "generalized point" or a "generalized vector" -- the distinction is mathematically unimportant. Thus we are free to describe the 5-tuple $(-2, 4, 0, 1, 6)$ either as a point in R^5 or a vector in R^5. We will use both descriptions.

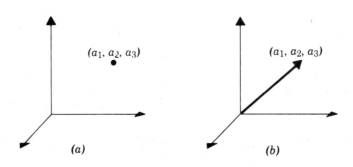

Figure 1

Definition. Two vectors $u = (u_1, u_2, \ldots, u_n)$ and $v = (v_1, v_2, \ldots, v_n)$ in R^n are called *equal* if

$$u_1 = v_1, \ u_2 = v_2, \ \ldots, \ u_n = v_n$$

The *sum* $u + v$ is defined by

$$u + v = (u_1 + v_1, \ u_2 + v_2, \ \ldots, \ u_n + v_n)$$

and if k is any scalar, the *scalar multiple* ku is defined by

$$ku = (ku_1, \ ku_2, \ \ldots, \ ku_n)$$

The operations of addition and scalar multiplication in this definition are called the *standard operations* on R^n.

We define the *zero vector* in R^n to be the vector

$$0 = (0, 0, \ldots, 0)$$

If $u = (u_1, u_2, \ldots, u_n)$ is any vector in R^n, then the *negative* (or *additive inverse*) of u is denoted by $-u$ and is defined by

$$-u = (-u_1, -u_2, \ldots, -u_n)$$

We define subtraction of vectors in R^n by $v - u = v + (-u)$, or in terms of components

$$v - u = v + (-u) = (v_1, v_2, \ldots, v_n) + (-u_1, -u_2, \ldots, -u_n)$$
$$= (v_1 - u_1, v_2 - u_2, \ldots, v_n - u_n)$$

The most important arithmetic properties of addition and scalar multiplication of vectors in R^n are listed in the next theorem.

Theorem 1. *If* $u = (u_1, u_2, \ldots, u_n)$, $v = (v_1, v_2, \ldots, v_n)$, *and* $w = (w_1, w_2, \ldots, w_n)$ *are vectors in* R^n *and* k *and* l *are scalars, then:*

(a) $u + v = v + u$

(b) $u + (v + w) = (u + v) + w$

(c) $u + 0 = 0 + u = u$

(d) $u + (-u) = 0$, *that is*, $u - u = 0$

(e) $k(lu) = (kl)u$

(f) $k(u + v) = ku + kv$

(g) $(k + l)u = ku + lu$

(h) $1u = u$

This theorem enables us to manipulate vectors in R^n without expressing the vectors in terms of components in much the same way that we manipulate real numbers. For example, to solve the vector equation $x + u = v$ for x, we can add $-u$ to both sides and proceed as follows.

$$(x + u) + (-u) = v + (-u)$$
$$x + (u - u) = v - u$$
$$x + 0 = v - u$$
$$x = v - u$$

The reader will find it useful to name parts of Theorem 1 that justify each of the steps in this computation.

To extend the notions of distance, norm, and angle to R^n, we begin with the following generalization of the dot product on R^2 and R^3.

Definition. If $u = (u_1, u_2, \ldots, u_n)$ and $v = (v_1, v_2, \ldots, v_n)$ are any vectors in R^n, then the *Euclidean inner product* $u \cdot v$ is defined by

$$u \cdot v = u_1 v_1 + u_2 v_2 + \cdots + u_n v_n$$

Observe that when $n = 2$ or 3, the Euclidean inner product is the ordinary dot product.

Example 1

The Euclidean inner product of the vectors

$$u = (-1, 3, 5, 7) \quad \text{and} \quad v = (5, -4, 7, 0)$$

in R^4 is

$$u \cdot v = (-1)(5) + (3)(-4) + (5)(7) + (7)(0) = 18$$

The four main arithmetic properties of the Euclidean inner product are listed in the next theorem.

Theorem 2. *If* **u**, **v**, *and* **w** *are vectors in* R^n *and* k *is any scalar, then:*

(a) $\mathbf{u} \cdot \mathbf{v} = \mathbf{v} \cdot \mathbf{u}$

(b) $(\mathbf{u} + \mathbf{v}) \cdot \mathbf{w} = \mathbf{u} \cdot \mathbf{w} + \mathbf{v} \cdot \mathbf{w}$

(c) $(k\mathbf{u}) \cdot \mathbf{v} = k(\mathbf{u} \cdot \mathbf{v})$

(d) $\mathbf{v} \cdot \mathbf{v} \geq 0$. *Further* $\mathbf{v} \cdot \mathbf{v} = 0$, *if and only if* $\mathbf{v} = 0$

Example 2

Theorem 2 allows us to perform computations with Euclidean inner products in much the same way that we perform them with ordinary arithmetic products. For example,

$$(3\mathbf{u} + 2\mathbf{v}) \cdot (4\mathbf{u} + \mathbf{v}) = (3\mathbf{u}) \cdot (4\mathbf{u} + \mathbf{v}) + (2\mathbf{v}) \cdot (4\mathbf{u} + \mathbf{v})$$
$$= (3\mathbf{u}) \cdot (4\mathbf{u}) + (3\mathbf{u}) \cdot \mathbf{v} + (2\mathbf{v}) \cdot (4\mathbf{u}) + (2\mathbf{v}) \cdot \mathbf{v}$$
$$= 12(\mathbf{u} \cdot \mathbf{u}) + 3(\mathbf{u} \cdot \mathbf{v}) + 8(\mathbf{v} \cdot \mathbf{u}) + 2(\mathbf{v} \cdot \mathbf{v})$$
$$= 12(\mathbf{u} \cdot \mathbf{u}) + 11(\mathbf{u} \cdot \mathbf{v}) + 2(\mathbf{v} \cdot \mathbf{v})$$

The reader should determine which parts of Theorem 2 were used in each step.

By analogy with the familiar formulas in R^2 and R^3, we define the *Euclidean norm* (or *Euclidean length*) of a vector $\mathbf{u} = (u_1, u_2, \ldots, u_n)$ in R^n by

$$\|\mathbf{u}\| = (\mathbf{u} \cdot \mathbf{u})^{1/2} = \sqrt{u_1^2 + u_2^2 + \cdots + u_n^2}$$

Similarly, the *Euclidean distance* between the points $\mathbf{u} = (u_1, u_2, \ldots, u_n)$ and $\mathbf{v} = (v_1, v_2, \ldots, v_n)$ in R^n is defined by

$$d(\mathbf{u}, \mathbf{v}) = \|\mathbf{u} - \mathbf{v}\| = \sqrt{(u_1 - v_1)^2 + (u_2 - u_2)^2 + \cdots + (u_n - v_n)^2}$$

Example 3

If $u = (1, 3, -2, 7)$ and $v = (0, 7, 2, 2)$ then

$$\|u\| = \sqrt{(1)^2 + (3)^2 + (-2)^2 + (7)^3} = \sqrt{63} = 3\sqrt{7}$$

and

$$d(u,v) = \sqrt{(1 - 0)^2 + (3 - 7)^2 + (-2 - 2)^2 + (7 - 2)^2} = \sqrt{58}$$

Since so many of the familiar ideas from 2-space and 3-space carry over, it is common to refer to R^n with the operations of addition, scalar multiplication, and inner product that we have defined here as *Euclidean n-space*.

We conclude this section by noting that it is possible to use the matrix notation

$$u = \begin{bmatrix} u_1 \\ u_2 \\ \vdots \\ u_n \end{bmatrix}$$

rather than the horizontal notation $u = (u_1, u_2, \ldots, u_n)$ to denote vectors in R^n. This is justified because the matrix operations

$$u + v = \begin{bmatrix} u_1 \\ u_2 \\ \vdots \\ u_n \end{bmatrix} + \begin{bmatrix} v_1 \\ v_2 \\ \vdots \\ v_n \end{bmatrix} = \begin{bmatrix} u_1 + v_1 \\ u_2 + v_2 \\ \vdots \\ u_n + v_n \end{bmatrix}$$

$$ku = k\begin{bmatrix} u_1 \\ u_2 \\ \vdots \\ u_n \end{bmatrix} = \begin{bmatrix} ku_1 \\ ku_2 \\ \vdots \\ ku_n \end{bmatrix}$$

produce the same results as the vector operations

$$u + v = (u_1, u_2, \ldots, u_n) + (v_1, v_2, \ldots, v_n) = (u_1 + v_1, u_2 + v_2, \ldots, u_n + v_n)$$

$$ku = k(u_1, u_2, \ldots, u_n) = (ku_1, ku_2, \ldots, ku_n)$$

The only difference is that results are displayed vertically in one case and horizontally in the other. We will use both notations.

Remark. It is common practice to omit the brackets on a 1 x 1 matrix. Thus, we might write 4 rather than [4]. Although this makes it impossible to tell whether 4 denotes the number "four" or the 1 x 1 matrix whose entry is "four", this rarely causes problems, since it is usually possible to tell which is meant from the context in which the symbol appears.

If we use matrix notation for the vectors

$$u = \begin{bmatrix} u_1 \\ u_2 \\ \vdots \\ u_n \end{bmatrix} \quad \text{and} \quad v = \begin{bmatrix} v_1 \\ v_2 \\ \vdots \\ v_n \end{bmatrix}$$

and omit the brackets on 1 x 1 matrices, then it follows that

$$v^t u = [v_1, v_2, \ldots, v_n] \begin{bmatrix} u_1 \\ u_2 \\ \vdots \\ u_n \end{bmatrix} = [u_1 v_1 + u_2 v_2 + \cdots + u_n v_n] = [u \cdot v] = u \cdot v$$

Thus, for vectors in vertical notation we have the matrix formula

$$v^t u = u \cdot v$$

for the Euclidean inner product. For example, if

$$u = \begin{bmatrix} -1 \\ 3 \\ 5 \\ 7 \end{bmatrix} \quad \text{and} \quad v = \begin{bmatrix} 5 \\ -4 \\ 7 \\ 0 \end{bmatrix}$$

then

$$u \cdot v = v^t u = \begin{bmatrix} 5 & -4 & 7 & 0 \end{bmatrix} \begin{bmatrix} -1 \\ 3 \\ 5 \\ 7 \end{bmatrix} = \begin{bmatrix} 18 \end{bmatrix} = 18$$

EXERCISE SET 3.1

1. Let $u = (2, 0, -1, 3)$, $v = (5, 4, 7, -1)$, and $w = (6, 2, 0, 9)$. Find:

    (a) $u - v$　　　　　　　　(b) $7v + 3w$　　　　　　　　(c) $-w + v$

    (d) $3(u - 7v)$　　　　　　(e) $-3v - 8w$　　　　　　　(f) $2v - (u + w)$

2. Let u, v, and w be the vectors in Exercise 1. Find the vector x that satisfies $2u - v + x = 7x + w$.

3. Let $u_1 = (-1, 3, 2, 0)$, $u_2 = (2, 0, 4, -1)$, $u_3 = (7, 1, 1, 4)$, and $u_4 = (6, 3, 1, 2)$. Find scalars c_1, c_2, c_3 and c_4 such that

    $c_1 u_1 + c_2 u_2 + c_3 u_3 + c_4 u_4 = (0, 5, 6, -3)$.

4. Show that there do not exist scalars c_1, c_2, and c_3 such that

    $c_1(1, 0, -2, 1) + c_2(2, 0, 1, 2) + c_3(1, -2, 2, 3) = (1, 0, 1, 0)$.

5. Compute the Euclidean norm of v when

    (a) $v = (4, -3)$　　　(b) $v = (1, -1, 3)$　　　(c) $v = (2, 0, 3, -1)$

    (d) $v = (-1, 1, 1, 3, 6)$

6. Let $u = (3, 0, 1, 2)$, $v = (-1, 2, 7, -3)$, and $w = (2, 0, 1, 1)$. Find
    (a) $\|u + v\|$ (b) $\|u\| + \|v\|$ (c) $\|-2u\| + 2\|u\|$

    (d) $\|3u - 5v + w\|$ (e) $\dfrac{1}{\|w\|} w$ (f) $\left\| \dfrac{1}{\|w\|} w \right\|$

7. Show that if v is a nonzero vector in R^n, then $(1/\|v\|)v$ has norm 1.

8. Find all scalars k such that $\|kv\| = 3$, where $v = (-1, 2, 0, 3)$.

9. Find the Euclidean inner product $u \cdot v$ when:
    (a) $u = (-1, 3)$, $v = (7, 2)$
    (b) $u = (3, 7, 1)$, $v = (-1, 0, 2)$
    (c) $u = (1, -1, 2, 3)$, $v = (3, 3, -6, 4)$
    (d) $u = (1, 3, 2, 6, -1)$, $v = (0, 0, 2, 4, 1)$

10. (a) Find two vectors in R^2 with Euclidean norm 1 whose Euclidean inner products with $(-1, 4)$ are zero.
    (b) Show that there are infinitely many vectors in R^3 with Euclidean norm 1 whose Euclidean inner product with $(-1, 7, 2)$ is zero.

11. Find the Euclidean distances between u and v when
    (a) $u = (2, -1)$, $v = (3, 2)$
    (b) $u = (1, 1, -1)$, $v = (2, 6, 0)$
    (c) $u = (2, 0, 1, 3)$, $v = (-1, 4, 6, 6)$
    (d) $u = (6, 0, 1, 3, 0)$, .$v = (-1, 4, 2, 8, 3)$

12. Establish the identity

$$\|u + v\|^2 + \|u - v\|^2 = 2\|u\|^2 + 2\|v\|^2$$

for vectors in R^n. Interpret this result geometrically in R^2.

13. Establish the identity

$$u \cdot v = \frac{1}{4} \|u + v\|^2 - \frac{1}{4} \|u - v\|^2$$

for vectors in R^n.

14. Verify parts (b), (e), (f) and (g) of Theorem 1 when $u = (1, 0, -1, 2)$, $v = (3, -1, 2, 4)$, $w = (2, 7, 3, 0)$, $k = 6$, and $\ell = -2$.

15. Verify parts (b) and (c) of Theorem 2 for the values of u, v, w, and k in Exercise 14.

16. Prove (a) through (d) of Theorem 1.

17. Prove (e) through (h) of Theorem 1.

18. Prove (a) and (c) of Theorem 2.

3.2 ROTATION OF AXES

In Section 12.5 of the text it was shown that if the axes of an xy-coordinate system are rotated about the origin through an angle θ to produce an x'y'-coordinate system (Figure 1)

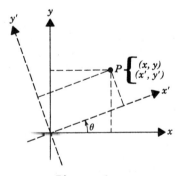

Figure 1

then the (x,y) and (x',y') coordinates of a point P are related by the *rotation equations*:

$$x = x' \cos \theta - y' \sin \theta$$
$$y = x' \sin \theta + y' \sin \theta$$

(1)

These equations can be written in matrix notation as

$$\begin{bmatrix} x \\ y \end{bmatrix} = \begin{bmatrix} \cos\theta & -\sin\theta \\ \sin\theta & \cos\theta \end{bmatrix} \begin{bmatrix} x' \\ y' \end{bmatrix} \tag{2}$$

The matrix

$$Q = \begin{bmatrix} \cos\theta & -\sin\theta \\ \sin\theta & \cos\theta \end{bmatrix}$$

is called the *rotation matrix* through the angle θ. We leave it for the reader to show that

$$Q^{-1} = \begin{bmatrix} \cos\theta & \sin\theta \\ -\sin\theta & \cos\theta \end{bmatrix}$$

Thus, (2) can be written

$$\begin{bmatrix} x' \\ y' \end{bmatrix} = \begin{bmatrix} \cos\theta & \sin\theta \\ -\sin\theta & \cos\theta \end{bmatrix} \begin{bmatrix} x \\ y \end{bmatrix} \tag{3}$$

Formula (2) is appropriate if the x'y'-coordinates are known and the xy-coordinates are to be computed, and Formula (3) is appropriate if the xy-coordinates are known and the x'y'-coordinates are to be computed. Also, Formula (2) is appropriate if an equation in xy-coordinates is to be transformed to an equation in x'y'-coordinates.

Example 1. Suppose that an x'y'-coordinate system is obtained by rotating an xy-coordinate system through an angle of $\theta = 45°$.

(a) Find the x'y'-coordinates of the point with xy-coordinates (2, -1).

(b) Find the xy-coordinates of the point with x'y'-coordinates (0,4).

(c) Find the equation of the curve xy = 1 in x'y'-coordinates.

Solution (a). From (3) with $\theta = 45°$, $x = 2$ and $y = -1$ we obtain

$$\begin{bmatrix} x' \\ y' \end{bmatrix} = \begin{bmatrix} \cos 45° & \sin 45° \\ -\sin 45° & \cos 45° \end{bmatrix} \begin{bmatrix} 2 \\ -1 \end{bmatrix}$$

$$= \begin{bmatrix} \dfrac{1}{\sqrt{2}} & \dfrac{1}{\sqrt{2}} \\ -\dfrac{1}{\sqrt{2}} & \dfrac{1}{\sqrt{2}} \end{bmatrix} \begin{bmatrix} 2 \\ -1 \end{bmatrix} = \begin{bmatrix} \dfrac{1}{\sqrt{2}} \\ -\dfrac{3}{\sqrt{2}} \end{bmatrix}$$

Thus, the x'y'-coordinates of the point are $\left(\dfrac{1}{\sqrt{2}}, -\dfrac{3}{\sqrt{2}} \right)$.

Solution (b). From (2) with $\theta = 45°$, $x' = 0$ and $y' = 4$ we obtain

$$\begin{bmatrix} x \\ y \end{bmatrix} = \begin{bmatrix} \cos 45° & -\sin 45° \\ \sin 45° & \cos 45° \end{bmatrix} \begin{bmatrix} 0 \\ 4 \end{bmatrix}$$

$$= \begin{bmatrix} \dfrac{1}{\sqrt{2}} & -\dfrac{1}{\sqrt{2}} \\ \dfrac{1}{\sqrt{2}} & \dfrac{1}{\sqrt{2}} \end{bmatrix} \begin{bmatrix} 0 \\ 4 \end{bmatrix} = \begin{bmatrix} -\dfrac{4}{\sqrt{2}} \\ \dfrac{4}{\sqrt{2}} \end{bmatrix}$$

Thus, the xy-coordinates of the point are $\left(-\dfrac{4}{\sqrt{2}}, \dfrac{4}{\sqrt{2}} \right)$.

Solution (c). From (2) with $\theta = 45°$, we obtain

$$\begin{bmatrix} x \\ y \end{bmatrix} = \begin{bmatrix} \cos 45° & -\sin 45° \\ \sin 45° & \cos 45° \end{bmatrix} \begin{bmatrix} x' \\ y' \end{bmatrix}$$

$$= \begin{bmatrix} \dfrac{1}{\sqrt{2}} & -\dfrac{1}{\sqrt{2}} \\ \dfrac{1}{\sqrt{2}} & \dfrac{1}{\sqrt{2}} \end{bmatrix} \begin{bmatrix} x' \\ y' \end{bmatrix} = \begin{bmatrix} \dfrac{x'}{\sqrt{2}} - \dfrac{y'}{\sqrt{2}} \\ \dfrac{x'}{\sqrt{2}} + \dfrac{y'}{\sqrt{2}} \end{bmatrix}$$

Thus,

$$x = \frac{x'}{\sqrt{2}} - \frac{y'}{\sqrt{2}}$$

$$y = \frac{x'}{\sqrt{2}} + \frac{y'}{\sqrt{2}}$$

Substituting these in the equation $xy = 1$ yields

$$\left(\frac{x'}{\sqrt{2}} - \frac{y'}{\sqrt{2}}\right)\left(\frac{x'}{\sqrt{2}} + \frac{y'}{\sqrt{2}}\right) = 1$$

or on simplifying

$$\frac{x'^2}{2} - \frac{y'^2}{2} = 1$$

We conclude this section by noting a more geometric method for deriving (3).

If an x'y'-coordinate system is obtained by rotating an xy-coordinate system through an angle θ, then the xy-coordinate system can be obtained by rotating the x'y'-coordinate system through the angle $-\theta$ (Figure 2).

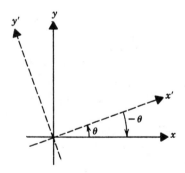

Figure 2

Thus, from (2) with the xy and x'y' roles reversed we obtain

$$\begin{bmatrix} x' \\ y' \end{bmatrix} = \begin{bmatrix} \cos(-\theta) & -\sin(-\theta) \\ \sin(-\theta) & \cos(-\theta) \end{bmatrix}\begin{bmatrix} x \\ y \end{bmatrix}$$

which reduces to (3) on applying the identities $\cos(-\theta) = \cos\theta$ and $\sin(-\theta) = -\sin\theta$.

EXERCISE SET 3.2

1. Find the matrix form of the rotation equations for a rotation of coordinate
    axes through

    (a) $0°$ (b) $30°$ (c) $45°$ (d) $60°$ (e) $90°$ (f) $180°$.

2. Let an x'y'-coordinate system be obtained by rotating an xy-coordinate system
    through an angle $\theta = 60°$.

    (a) Find the rotation matrix through the angle θ.

    (b) Find the inverse of the rotation matrix in (a).

    (c) Use (3) to find the x'y'-coordinates of the point whose xy-coordinates
        are (1, 5).

    (d) Use (2) to find the xy-coordinates of the point whose x'y'-coordinates
        are (-1, 5).

    (e) Find the equation of the curve $\sqrt{3}xy + y^2 = 2$ in x'y'-coordinates.

3. Let an x'y'-coordinate system be obtained by rotating an xy-coordinate system
    through an angle $\theta = 30°$.

    (a) Find the rotation matrix through the angle θ.

    (b) Find the inverse of the rotation matrix in (a).

    (c) Use (3) to find the x'y'-coordinates of the point whose xy-coordinates
        are (-2, 4).

    (d) Use (2) to find the xy-coordinates of the point whose x'y'-coordinates
        are (6, -5).

    (e) Find the equation of the curve $x^2 + \sqrt{3}xy - 5 = 0$ in x'y'-coordinates.

4. Let an x'y'-coordinate system be obtained by rotating an xy-coordinate system
    through an angle $\theta = 45°$.

    (a) Find the rotation matrix through the angle θ.

    (b) Find the inverse of the rotation matrix in (a).

    (c) Use (2) to find the x'y'-coordinates of the point whose xy-coordinates
        are (4, 1).

    (d) Use (3) to find the xy-coordinates of the point whose x'y'-coordinates
        are (-6, 0).

    (e) Find the equation of the curve $x^2 - xy + y^2 = 1$ in x'y'-coordinates.

5. Let an x'y'-coordinate system be obtained by rotating an xy-coordinate system through an angle $\theta = 45°$. Find the equation in xy-coordinates of the curve $x'^2 + 3y'^2 = 5$.

6. (a) Let an x'y'z'-coordinate system be obtained by rotating an xyz-coordinate system around its z-axis counterclockwise (looking along the positive z-axis toward the origin) through an angle θ. Find a matrix Q such that

$$\begin{bmatrix} x \\ y \\ z \end{bmatrix} = Q \begin{bmatrix} x' \\ y' \\ z' \end{bmatrix}$$

   where (x, y, z) and (x', y', z') are coordinates of the same point in the xyz-system and x'y'z'-system respectively.

   (b) Repeat (a) for a rotation about the x-axis.

   (c) Repeat (a) for a rotation about the y-axis.

7. A triangle has vertices (-1, 0), (0, 1) and (1, 0) in an xy-coordinate. Find the new coordinates of the vertices if the xy-axes are rotated through $-135°$.

8. An xy-coordinate system is rotated through an angle θ to obtain x'y'-coordinate system. Find the equation of the line x = 0 in x'y'-coordinates.

3.3 TRANSFORMATIONS FROM R^n TO R^m

In Section 2.1 we defined a function to be a rule that assigns to each element in a set A one and only one element in a set B. In our work thus far, A and B have been sets in R, R^2 and R^3 primarily. Some examples are listed in the following table:

Table 1

Formula	Example	Classification	Description
$f(x)$	$f(x) = x^2$	Real valued function of a real variable	Function from R to R
$f(x,y)$	$f(x,y) = x^2 + y^2$	Real valued function of two real variables	Function from R^2 to R
$f(x,y;z)$	$f(x,y,z)$ $= x^2 + y^2 + z^2$	Real valued function of three real variables	Function from R^3 to R
$r(t) = x(t)i + y(t)j$	$r(t) = ti + t^2 j$	Vector valued function of a real variable (two parametric equations)	Function from R to R^2
$r(t) = x(t)i + y(t)j + z(t)k$	$r(t) = \sin ti + \cos tj + tk$	Vector valued function of a real variable (three parametric equations)	Function from R to R^3

More generally, we shall be interested in functions from R^n to R^m. However, before we discuss such functions it will be helpful to introduce some terminology and notation.

If the domain of a function f is a subset of R^n and the range is a subset of R^m, then f is called a *transformation* from R^n to R^m and we say that f *maps* R^n to R^m. We denote this by writing $f:R^n \to R^m$.

Example 1

The function $f(x,y) = xy + 1$ assigns a real number to each point (x,y) in R^2, so f is a transformation from R^2 to R. We can say that f maps R^2 to R and write $f:R^2 \to R$.

The following examples illustrate various ways in which functions from R^n to R^m can be defined.

Example 2

The function

$$f(x_1, x_2, \ldots, x_n) = x_1^2 + x_2^2 + \ldots + x_n^2$$

assigns a real number to each point $(x_1, x_2, \ldots, x_n)$ in R^n, so f is a transformation from R^n to R.

Example 3

Let

$$x_1 = x_1(t), \ x_2 = x_2(t), \ \ldots, \ x_m = x_m(t)$$

be m parametric equations and for each value of t construct the vector $r(t)$ with $x_1(t), x_2(t), \ldots, x_m(t)$ as components; that is

$$r(t) = (x_1(t), x_2(t), \ldots, x_m(t))$$

The function $r(t)$ assigns a unique point (vector) in R^m to each value of t. So r is a transformation from R to R^m. For example,

$$r(t) = (t, t^2, t^3, t^4)$$

defines a transformation from R to R^4.

Example 4

Let

$$y_1 = f_1(x_1, x_2, \ldots, x_n)$$
$$y_2 = f_2(x_1, x_2, \ldots, x_n)$$
$$\vdots$$
$$y_m = f_m(x_1, x_2, \ldots, x_n)$$

where f_1, f_2, ..., f_m are real valued functions of the real variables x_1, x_2, ..., x_n. These m equations assign a unique point $(y_1, y_2, ..., y_m)$ in R^m to each point $(x_1, x_2, ..., x_n)$ in R^n. Thus, the m equations define a transformation from R^n to R^m. If we denote this transformation by T, then $T:R^n \rightarrow R^m$ and

$$T(x_1, x_2, ..., x_n) = (y_1, y_2, ..., y_m)$$

or equivalently

$$T(x_1, x_2, ..., x_n) = (f_1(x_1, x_2, ..., x_n), f_2(x_1, x_2, ..., x_n), ..., f_m(x_1, x_2, ..., x_n))$$

Example 5

As an illustration of the general result in Example 4, suppose that

$$y_1 = x_1 + x_2$$
$$y_2 = 3x_1 x_2$$
$$y_3 = x_1^2 - x_2^2$$

These equations define a transformation $T:R^2 \rightarrow R^3$ given by the formula

$$T(x_1, x_2) = \left(x_1 + x_2, 3x_1 x_2, x_1^2 - x_2^2\right)$$

Thus, for example

$$T(1, -2) = (-1, -6, -3)$$

Example 6

For each point $(x_1, x_2, ..., x_n)$ in R^n, the matrix equation

$$\begin{bmatrix} y_1 \\ y_2 \\ \vdots \\ y_m \end{bmatrix} = \begin{bmatrix} a_{11} & a_{12} & \cdots & a_{1n} \\ a_{21} & a_{22} & \cdots & a_{2n} \\ \vdots & \vdots & & \vdots \\ a_{m1} & a_{m2} & \cdots & a_{mn} \end{bmatrix} \begin{bmatrix} x_1 \\ x_2 \\ \vdots \\ x_n \end{bmatrix}$$

determines a unique point $(y_1, y_2, \ldots, y_m)$ in R^m. Thus, this matrix equation defines a transformation from R^n to R^m. Transformations of this kind are of special importance and are studied extensively in linear algebra.

In the remainder of this section we shall focus our attention on transformations from R^2 to R^2 and R^3 to R^3.

If the transformation $T:R^2 \to R^2$ is given by the equations

$$u = \phi(x,y), \quad v = \psi(x,y) \tag{1}$$

then T can be pictured geometrically as in Figure 1, where we show the transformation T mapping a point (x,y) in the xy-plane to the point (u,v) in the uv-plane.

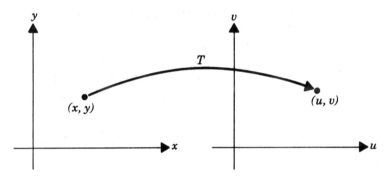

Figure 1

One way to describe the geometric effect of T is to sketch the sets of points in the xy-plane that map into vertical and horizontal lines in the uv-plane. Sets of points that map into vertical lines are called u-*curves* and sets that map into horizontal lines are called v-*curves* (Figure 2).

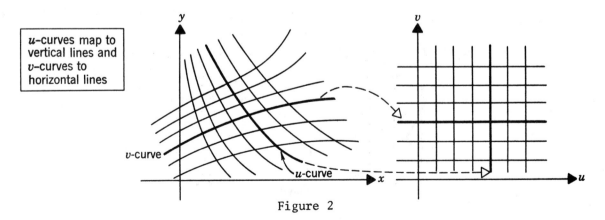

u-curves map to vertical lines and v-curves to horizontal lines

Figure 2

Example 4

Let $T:R^2 \to R^2$ be the transformation defined by

$$u = 2x + y, \qquad v = 2x - y$$

(a) Find $T(3,1)$, $T(0,0)$, and $T(-1,4)$.
(b) Sketch the u-curves that map into $u = 2$ and $u = -2$.
(c) Sketch the v-curves that map into $v = 2$ and $v = -2$.

Solution (a). Substituting in the formula

$$T(x,y) = (u,v) = (2x + y, \ 2x - y)$$

yields

$$T(3,1) = (7,5), \ T(0,0) = (0,0), \ T(-1, \ 4) = (2, \ -6)$$

Solution (b). Since $u = 2x + y$ and $v = 2x - y$ the u-curves that map into $u = 2$ and $u = -2$ respectively are the lines

$$2x + y = 2, \qquad 2x + y = -2$$

and the v-curves that map into $v = 2$ and $v = -2$ are the lines

$$2x - y = 2, \qquad 2x - y = -2$$

Solution (c)

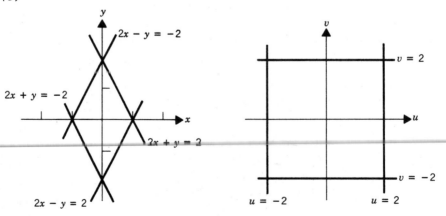

Figure 3

If a transformation $T:R^2 \to R^2$ is defined by the formulas

$$u = \phi(x,y), \quad v = \psi(x,y)$$

and if there is no mention of the domain of T, then it is understood that the domain consists of all points in the xy-plane at which both $\phi(x,y)$ and $\psi(x,y)$ make sense and yield real values. Stated another way, the domain of T is the intersection of domains of ϕ and ψ.

Recall that if f is a real valued function of a real variable, then the function f is called one-to-one (1-1) if it does not assign the same value to any two distinct points in its domain (Definition 8.1.2). Similarly, we define T to be *one-to-one*, written 1-1, if the transformation T assigns distinct points in the uv-plane to distinct points in its domain.

Example 5

Let T be the transformation defined by

$$u = 2x + y, \quad v = 2x - y$$

The functions $f(x,y) = 2x + y$ and $g(x,y) = 2x - y$ are defined at all points in the xy-plane and so the domain of T is the entire xy-plane. We leave it as an exercise to show that T is 1-1 (Exercise 9).

If a transformation T, defined by

$$u = \phi(x,y), \quad v = \psi(x,y) \tag{2}$$

is one-to-one then the "reverse" transformation that maps the point (u,v) in the uv-plane to the point (x,y) in the xy-plane is called the *inverse* of T and is denoted by T^{-1}. Thus,

$$T(x,y) = (u,v) \quad \text{and} \quad T^{-1}(u,v) = (x,y)$$

A formula for T^{-1} is given by equations of the form

$$x = f(u,v), \quad y = g(u,v)$$

In some cases these equations can be obtained by solving (2) for x and y in terms of u and v.

Example 6

Find a formula for T^{-1} if T is given by

$$u = 2x + y, \qquad v = 2x - y$$

Solution. We will solve the linear system

$$2x + y = u$$

$$2x - y = v$$

for x and y in terms of u and v. Using Cramer's rule we obtain

$$x = \frac{\begin{vmatrix} u & 1 \\ v & -1 \end{vmatrix}}{\begin{vmatrix} 2 & 1 \\ 2 & -1 \end{vmatrix}} = \frac{-u - v}{-4} = \frac{1}{4}(u + v)$$

$$y = \frac{\begin{vmatrix} 2 & u \\ 2 & v \end{vmatrix}}{\begin{vmatrix} 2 & 1 \\ 2 & -1 \end{vmatrix}} = \frac{2v - 2u}{-4} = \frac{1}{2}(u - v)$$

Thus, $T^{-1}(u,v) = \left(\frac{1}{4} u + \frac{1}{4} v, \frac{1}{2} u - \frac{1}{2} v \right)$.

Remark. We were able to apply Cramer's rule in this example only because the equations in the system are linear. If the equations are not linear, then other methods must be devised to solve for x and y. This can be difficult or even impossible in some cases.

Let P(x,y) be any point in the xy-plane and let (r,θ) be the polar coordinates of P that satisfy

$$r \geq 0 \quad \text{and} \quad 0 \leq \theta < 2\pi$$

(For the origin, take the polar coordinates to be r = 0 and θ = 0.) The transformation T is defined by

$$T(x,y) = (r,\theta)$$

maps the xy-plane into the rθ-plane and is called the *transformation from rectangular coordinates to polar coordinates*. It can be shown that this transformation is 1-1 (Exercise 18).

Example 7

Let T(x,y) = (r,θ) be the transformation from rectangular coordinate to polar coordinates.

(a) Sketch the graphs of r = 1, r = 2, r = 3, and θ = 0, θ = π/2, and θ = π in the rθ-plane.

(b) Sketch the r-curves in the xy-plane that map into r = 1, r = 2, and r = 3.

(c) Sketch the θ-curves in the xy-plane that map into θ = π/4, θ = π/2, and θ = 3π/4.

(d) Find a formula for T^{-1}.

Solution (a).   The graphs are the lines shown in Figure 4.

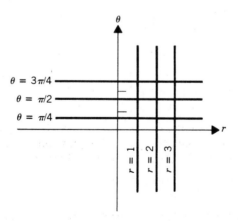

Figure 4

Solution (b). The r-curves in the xy-plane are the circles with polar equations r = 1, r = 2, and r = 3 (Figure 5).

Solution (c). The θ-curves in the xy-plane are the rays with equations θ = 0, θ = π/2, and θ = π (Figure 5)

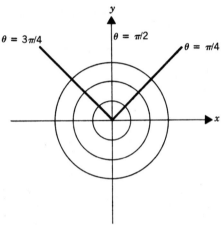

Figure 5

Solution (d). The equations that express rectangular coordinates (x,y) in terms of polar coordinates (r,θ) are

$$x = r\cos\theta, \quad y = r\sin\theta$$

so the inverse of the polar transformation T is given by

$$T^{-1}(r,\theta) = (r\cos\theta, \; r\sin\theta)$$

This is called the *transformation from polar to rectangular coordinates.*

The concepts discussed in this section have natural extensions to three dimensions. For example, if the transformation $T:R^3 \to R^3$ is defined by the equations

$$u = \phi(x,y,z), \quad v = \psi(x,y,z), \quad w = \lambda(x,y,z)$$

then T maps a point in xyz-space into a point (u,v,w,) in uvw-space (Figure 6). Those points in xyz-space that map into planes of the form

$$u = constant, \quad v = constant, \quad w = constant$$

are called *u-surfaces, v-surfaces,* and *w-surfaces,* respectively. We leave it for the reader to define the terms, *domain, one-to-one,* and *inverse* for such transformations.

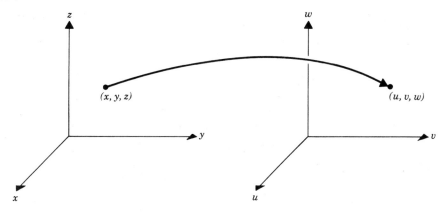

Figure 6

Exercise Set 3.3

1. In each part the formulas define a transformation from R^n to R^m. Find m and n.

    (a) $f(x,y) = x \, e^y$

    (b) $r(t) = t^3 i - t^2 j$

    (c) $f(x_1, x_2, x_3) = \sqrt{x_1^2 + x_2^2 + x_3^2}$

    (d) $r(t) = (t, -2t, 0, t^3)$

    (e) $y_1 = x_1 - x_2 - x_3$

    $\quad\; y_2 = 3x_1 - x_3$

    (f) $\begin{bmatrix} y_1 \\ y_2 \\ y_3 \end{bmatrix} = \begin{bmatrix} 1 & 2 \\ -1 & 3 \\ 5 & 7 \end{bmatrix} \begin{bmatrix} x_1 \\ x_2 \end{bmatrix}$

    (g) $T(x_1, x_2, x_3) = (x_2, -x_1, x_3)$

    (h) $u = x \sin y + y$

    $\quad\; v = xy$

2. Let T be the transformation defined by the formulas

    $$y_1 = x_1 x_2 \; , \; y_2 = x_1^2 + x_2^2 \; , \; y_3 = x_1 - x_2 \; .$$

    Find

    (a) $T(0, 0)$

    (b) $T(-1, 2)$

    (c) $T(a, b)$.

3. Let $T : R^2 \rightarrow R^2$ be the transformation defined by

    $$u = x - 2y \; , \quad v = y + 2x$$

    (a) Find $T(-1, 3)$, $T(2, 6)$, $T(0, 0)$.

    (b) Sketch the u-curves that map into $u = -3$, $u = 0$, $u = 3$.

    (c) Sketch the v-curves that map into $v = -3$, $v = 0$, $v = 3$.

4. Let $T : R^2 \rightarrow R^2$ be the transformation defined by

    $$u = xy \; , \quad v = x^2 - y^2$$

    (a) Find $T(2, 4)$, $T(-1, -3)$, $T(0,5)$

(b) Sketch the u-curves that map into u = 1, u = -1, u = 4, u = -4.

(c) Sketch the v-curves that map into v = 1, v = -1, v = 4, v = -4.

In Exercises 5-8, sketch the domain of the transformation $T : R^2 \to R^2$ defined by the formulas.

5. $u = x + y$, $v = \sqrt{xy}$

6. $u = \dfrac{x}{x^2 + y^2}$, $v = \dfrac{y}{x^2 + y^2}$

7. $u = \dfrac{1}{1 - x - y}$, $v = \dfrac{1}{1 + x + y}$

8. $u = x^2 + y^2$, $v = x^2 - y^2$

9. Show that the transformation T defined by u = 2x + y, v = 2x - y is one-to-one.

In Exercises 10-15, let T be the transformation defined by the given formulas. Determine whether T is one-to-one and if so find formulas x = f(u, v) , y = g(u, v) for T^{-1}.

10. u = 3x + y , v = x - y

11. u = x + y , v = xy

12. $u = x^2 - y^2$, $v = x^2 + y^2$

13. $u = x^3$, v = x - y

14. Find a transformation x = f(u,v), y = g(u,v) which maps the ellipse

$$\frac{x^2}{a^2} + \frac{y^2}{b^2} = 1 \quad \text{onto the unit circle } u^2 + v^2 = 1.$$

15. Let T be the transformation defined by u = x + 2y , v = 2x - 3y

(a) Sketch the image in the uv-plane of the rectangle in the xy-plane with vertices (0,0), (1,0), (1,1), (0,1).

(b) Find an equation for the image of the line 4x + 2y = 5.

16. For the transformation in Exercise 15, find an equation for the curve in the xy-plane that maps into the line u + v = 5.

17. For the transformation in Exercise 15, sketch the image in the uv-plane enclosed by x + y = 1, y - x = 1, y = 0.

18. Show that the transformation from rectangular coordinates (x,y) to polar coordinates (r,θ), where r ≥ 0, 0 ≤ θ < 2π is one-to-one.

19. Let $T(x,y) = (r,\theta)$ be the transformation from rectangular coordinates to polar coordinates.

    (a) Sketch the r-curves in the xy-plane that map into $r = 4$, $r = 5$, and $r = 6$.

    (b) Sketch the θ-curves in the xy-plane that map into $\theta = \pi/4$, $\theta = \pi/6$, $\theta = \pi/3$.

20. Let $P(x, y, z)$ be any point in xyz-space and (r, θ, z) the cylindrical coordinates of P, where $r \geq 0$, $0 \leq \theta < 2\pi$. (For the origin, take the cylindrical coordinates to be $r = 0$, $\theta = 0$, $z = 0$.) Then the transformation $T(x, y, z) = (r, \theta, z)$ defines a mapping from xyz-space to $r\theta z$-space, called the *transformation from rectangular to cylindrical coordinates*.

    (a) Describe the r-surface $r = 1$.

    (b) Describe the θ-surface $\theta = \pi/4$.

    (c) Describe the z-surface $z = 1$.

21. Let $P(x,y,z)$ be any point in xyz-space and let (ρ, θ, ϕ) be the spherical coordinates of P, where $\rho \geq 0$, $0 \leq \theta < 2\pi$, $0 \leq \phi \leq \pi$. (For the origin, take the spherical coordinates to be $\rho = 0$, $\theta = 0$, $\phi = 0$.) Then the transformation $T(x, y, z) = (\rho, \theta, \phi)$ defines a mapping from xyz-space to $\rho\theta\phi$-space, called the transformation from *rectangular to spherical coordinates*.

    (a) Describe the ρ-surface $\rho = 1$.

    (b) Describe the θ-surface $\theta = \pi/4$.

    (c) Describe the ϕ-surface $\phi = \pi/4$.

ANSWERS TO EXERCISES

CHAPTER 1

Exercise Set 1.1 (page 8)

1. b, d, f

2. (a) $x = \frac{7}{6} t + \frac{1}{2}$, $y = t$

(b) $x_1 = -2s + \frac{7}{2} t + 4$, $x_2 = s$, $x_3 = t$

(c) $x_1 = \frac{4}{3} r - \frac{7}{3} s + \frac{8}{3} t - \frac{5}{3}$, $x_2 = r$, $x_3 = s$, $x_4 = t$

(d) $v = \frac{1}{2} q - \frac{3}{2} r - \frac{1}{2} s + 2t$, $w = q$, $x = r$, $y = s$, $z = t$

3. (a) $\begin{bmatrix} 1 & -2 & 0 \\ 3 & 4 & -1 \\ 2 & -1 & 3 \end{bmatrix}$

(b) $\begin{bmatrix} 1 & 0 & 1 & 1 \\ -1 & 2 & -1 & 3 \end{bmatrix}$

(c) $\begin{bmatrix} 1 & 0 & 1 & 0 & 0 & 1 \\ 0 & 2 & -1 & 0 & 1 & 2 \\ 0 & 0 & 2 & 1 & 0 & 3 \end{bmatrix}$

(d) $\begin{bmatrix} 1 & 0 & 1 \\ 0 & 1 & 2 \end{bmatrix}$

4. (a) $\begin{aligned} x_1 \quad - x_3 &= 2 \\ 2x_1 + x_2 + x_3 &= 3 \\ - x_2 + 2x_3 &= 4 \end{aligned}$

(b) $\begin{aligned} x_1 \quad &= 0 \\ x_2 &= 0 \\ x_1 - x_2 &= 1 \end{aligned}$

(c) $\begin{aligned} x_1 + 2x_2 + 3x_3 + 4x_4 &= 5 \\ 5x_1 + 4x_2 + 3x_3 + 2x_4 &= 1 \end{aligned}$

(d) $\begin{aligned} x_1 \quad &= 1 \\ x_2 &= 2 \\ x_3 &= 3 \\ x_4 &= 4 \end{aligned}$

5. $k = 6$ infinitely many solutions

    $k \neq 6$ no solutions

6. (a) The lines have no common point of intersection.

    (b) The lines intersect in exactly one point.

    (c) The three lines coincide.

Exercise Set 1.2 (page 19)

1. d, f

2. b, c, f

3. $x_1 = 4$, $x_2 = 3$, $x_3 = 2$

4. $x_1 = 2 - 3t$, $x_2 = 4 + t$, $x_3 = 2 - t$, $x_4 = t$

5. $x_1 = -1 - 5s - 5t$, $x_2 = s$, $x_3 = 1 - 3t$, $x_4 = 2 - 4t$, $x_5 = t$

6. Inconsistent

7. $x_1 = 3$, $x_2 = 1$, $x_3 = 2$

8. $x_1 = -3t$, $x_2 = -4t$, $x_3 = 7t$

9. $x = s - 1$, $y = 2r$, $z = r$, $w = s$

10. Inconsistent

11. $x_1 = -4$, $x_2 = 2$, $x_3 = 7$

12. $x_1 = 3 + 2t$, $x_2 = t$

13. $x_1 = 0$, $x_2 = -3t$, $x_3 = t$

14. Inconsistent

15. (a) $x_1 = \frac{2}{3} a - \frac{1}{9} b$, $x_2 = -\frac{1}{3} a + \frac{2}{9} b$

    (b) $x_1 = a - \frac{1}{3} c$, $x_2 = a - \frac{1}{2} b$, $x_3 = -a + \frac{1}{2} b + \frac{1}{3} c$

16. $a = 4$ infinitely many, $a = -4$ none, $a \neq \pm 4$ exactly one.

18. $\begin{bmatrix} 1 & 3 \\ 0 & 1 \end{bmatrix}$ and $\begin{bmatrix} 1 & 0 \\ 0 & 1 \end{bmatrix}$ are possible answers.

19. $\alpha = \frac{\pi}{2}$, $\beta = \pi$, $\gamma = 0$.

Exercise Set 1.3 (page 25)

1. a, c, d

2. $x_1 = 0$, $x_2 = 0$, $x_3 = 0$

3. $x_1 - \frac{1}{4} s$, $x_2 = -\frac{1}{4} s - t$, $x_3 = s$, $x_4 = t$

4. $x_1 = 0$, $x_2 = 0$, $x_3 = 0$, $x_4 = 0$

5. $x = \frac{t}{8}$, $y = \frac{5t}{16}$, $z = t$

6. $\lambda = 4$, $\lambda = 2$

11. One possible answer is $x + y + z = 0$
$$x + y + z = 1$$

Exercise Set 1.4 (page 35)

1. (a) Undefined (b) 4×2 (c) Undefined
   (d) Undefined (e) 5×5 (f) 5×2

3. $a = 5$, $b = -3$, $c = 4$, $d = 1$

4. (a) $\begin{bmatrix} 12 & -3 \\ -4 & 5 \\ 4 & 1 \end{bmatrix}$ (b) $\begin{bmatrix} 7 & 6 & 5 \\ -2 & 1 & 3 \\ 7 & 3 & 7 \end{bmatrix}$ (c) $\begin{bmatrix} -5 & 4 & -1 \\ 0 & -1 & -1 \\ -1 & 1 & 1 \end{bmatrix}$

(d) $\begin{bmatrix} 9 & 8 & 19 \\ -2 & 0 & 0 \\ 32 & 9 & 25 \end{bmatrix}$ (e) $\begin{bmatrix} 14 & 36 & 25 \\ 4 & -1 & 7 \\ 12 & 26 & 21 \end{bmatrix}$ (f) $\begin{bmatrix} -28 & 7 \\ 0 & -14 \end{bmatrix}$

(g) $\begin{bmatrix} 4 & 3 & 3 \\ 3 & 3 & 6 \end{bmatrix}$ (h) $\begin{bmatrix} -4 & -1 & 2 \\ 9 & -1 & 3 \\ 1 & 0 & 5 \end{bmatrix}$

5. (a) Undefined (b) $\begin{bmatrix} 42 & 108 & 75 \\ 12 & -3 & 21 \\ 36 & 78 & 63 \end{bmatrix}$ (c) $\begin{bmatrix} 3 & 45 & 9 \\ 11 & -11 & 17 \\ 7 & 17 & 13 \end{bmatrix}$

(d) $\begin{bmatrix} 3 & 45 & 9 \\ 11 & -11 & 17 \\ 7 & 17 & 13 \end{bmatrix}$ (e) Undefined (f) $\begin{bmatrix} 48 & 15 & 31 \\ 0 & 2 & 6 \\ 38 & 10 & 27 \end{bmatrix}$

(g) Undefined (h) $\begin{bmatrix} 4 & 5 \\ 16 & -2 \\ 8 & 8 \end{bmatrix}$ (i) $\begin{bmatrix} 12 & -4 & 4 \\ -3 & 5 & 1 \end{bmatrix}$

6. (a) $\begin{bmatrix} 67 & 41 & 41 \end{bmatrix}$ (b) $\begin{bmatrix} 63 & 67 & 57 \end{bmatrix}$ (c) $\begin{bmatrix} 41 \\ 21 \\ 67 \end{bmatrix}$

(d) $\begin{bmatrix} 6 \\ 6 \\ 63 \end{bmatrix}$ (e) $\begin{bmatrix} 24 & 56 & 97 \end{bmatrix}$ (f) $\begin{bmatrix} 76 \\ 98 \\ 97 \end{bmatrix}$

7. 182

Exercise Set 1.5 (page 48)

3. $A^{-1} = \begin{bmatrix} 2 & -1 \\ -5 & 3 \end{bmatrix}$ $B^{-1} = \begin{bmatrix} \frac{1}{5} & \frac{3}{20} \\ -\frac{1}{5} & \frac{1}{10} \end{bmatrix}$ $C^{-1} = \begin{bmatrix} \frac{1}{2} & 0 \\ 0 & \frac{1}{3} \end{bmatrix}$

5. No

6. $\begin{bmatrix} -3 & 2 \\ \frac{5}{2} & -\frac{3}{2} \end{bmatrix}$

7. $\begin{bmatrix} 1 & \frac{2}{7} \\ \frac{4}{7} & \frac{1}{7} \end{bmatrix}$

8. $A^3 = \begin{bmatrix} 1 & 0 \\ 26 & 27 \end{bmatrix}$ $A^{-3} = \begin{bmatrix} 1 & 0 \\ -\frac{26}{27} & \frac{1}{27} \end{bmatrix}$ $A^2 - 2A + I = \begin{bmatrix} 0 & 0 \\ 4 & 4 \end{bmatrix}$

9. $A^{-1} = \begin{bmatrix} \frac{1}{2} & -\frac{1}{2} & \frac{1}{2} \\ \frac{1}{2} & \frac{1}{2} & -\frac{1}{2} \\ -\frac{1}{2} & \frac{1}{2} & \frac{1}{2} \end{bmatrix}$

10. $\begin{bmatrix} \cos\theta & -\sin\theta \\ \sin\theta & \cos\theta \end{bmatrix}$

11. (c) $(A + B)^2 = A^2 + AB + BA + B^2$

12. $A^{-1} = \begin{bmatrix} \frac{1}{a_{11}} & 0 & \cdots & 0 \\ 0 & \frac{1}{a_{22}} & \cdots & 0 \\ \vdots & \vdots & & \vdots \\ 0 & 0 & \cdots & \frac{1}{a_{nn}} \end{bmatrix}$

21. 0A and A0 may not have the same size.

22. $\begin{bmatrix} \pm1 & 0 & 0 \\ 0 & \pm1 & 0 \\ 0 & 0 & \pm1 \end{bmatrix}$

Exercise Set 1.6 (page 55)

1. $\begin{bmatrix} -5 & 2 \\ 3 & -1 \end{bmatrix}$

2. $\begin{bmatrix} -5 & -3 \\ -3 & -2 \end{bmatrix}$

3. Not Invertible.

4. $\begin{bmatrix} \dfrac{3}{2} & -\dfrac{11}{10} & -\dfrac{6}{5} \\ -1 & 1 & 1 \\ -\dfrac{1}{2} & \dfrac{7}{10} & \dfrac{2}{5} \end{bmatrix}$

5. Not Invertible.

6. $\begin{bmatrix} \dfrac{1}{2} & -\dfrac{1}{2} & \dfrac{1}{2} \\ -\dfrac{1}{2} & \dfrac{1}{2} & \dfrac{1}{2} \\ \dfrac{1}{2} & \dfrac{1}{2} & -\dfrac{1}{2} \end{bmatrix}$

7. $\begin{bmatrix} 7/2 & 0 & -3 \\ -1 & 1 & 0 \\ 0 & -1 & 1 \end{bmatrix}$

8. $\begin{bmatrix} \dfrac{1}{2} & -\dfrac{1}{2} & \dfrac{1}{2} \\ 0 & 0 & 1 \\ \dfrac{1}{2} & \dfrac{1}{2} & -\dfrac{1}{2} \end{bmatrix}$

9. $\begin{bmatrix} 1 & 0 & -2 \\ 3 & 1 & 2 \\ 1 & -1 & 0 \end{bmatrix}$

10. $\begin{bmatrix} \dfrac{1}{2}\sqrt{2} & -\dfrac{1}{2}\sqrt{2} & 0 \\ \dfrac{1}{2}\sqrt{2} & \dfrac{1}{2}\sqrt{2} & 0 \\ 0 & 0 & 1 \end{bmatrix}$

11. $\begin{bmatrix} 1 & 0 & 0 & 0 \\ -\dfrac{1}{2} & \dfrac{1}{2} & 0 & 0 \\ 0 & -\dfrac{1}{4} & \dfrac{1}{4} & 0 \\ 0 & 0 & -\dfrac{1}{8} & \dfrac{1}{8} \end{bmatrix}$

12. Not Invertible

13. $A^{-1} = \begin{bmatrix} \cos\theta & -\sin\theta & 0 \\ \sin\theta & \cos\theta & 0 \\ 0 & 0 & 1 \end{bmatrix}$

14. (a) $\begin{bmatrix} \dfrac{1}{k_1} & 0 & 0 & 0 \\ 0 & \dfrac{1}{k_2} & 0 & 0 \\ 0 & 0 & \dfrac{1}{k_3} & 0 \\ 0 & 0 & 0 & \dfrac{1}{k_4} \end{bmatrix}$
 (b) $\begin{bmatrix} 0 & 0 & 0 & \dfrac{1}{k_4} \\ 0 & 0 & \dfrac{1}{k_3} & 0 \\ 0 & \dfrac{1}{k_2} & 0 & 0 \\ \dfrac{1}{k_1} & 0 & 0 & 0 \end{bmatrix}$

(c) $\begin{bmatrix} \dfrac{1}{k} & 0 & 0 & 0 \\ -\dfrac{1}{k^2} & \dfrac{1}{k} & 0 & 0 \\ \dfrac{1}{k^3} & -\dfrac{1}{k^2} & \dfrac{1}{k} & 0 \\ -\dfrac{1}{k^4} & \dfrac{1}{k^3} & -\dfrac{1}{k^2} & \dfrac{1}{k} \end{bmatrix}$

Exercise Set 1.7 (page 59)

1. $x_1 = 41$, $x_2 = -17$

2. $x_1 = \dfrac{46}{27}$, $x_2 = -\dfrac{13}{27}$

3. $x_1 = -7$, $x_2 = 4$, $x_3 = -1$

4. $x_1 = 1$, $x_2 = -11$, $x_3 = 16$

5. $x = 1$, $y = 5$, $z = -1$

6. $w = 1$, $x = -6$, $y = 10$, $z = -7$

7. (a) $x_1 = \frac{16}{3}$, $x_2 = -\frac{4}{3}$, $x_3 = -\frac{11}{3}$ (b) $x_1 = -\frac{5}{3}$, $x_2 = \frac{5}{3}$, $x_3 = \frac{10}{3}$

     (c) $x_1 = 3$, $x_2 = 0$, $x_3 = -4$ (d) $x_1 = \frac{41}{42}$, $x_2 = -\frac{5}{6}$, $x_3 = \frac{25}{21}$

8. (a) $b_2 = 3b_1$, $b_3 = -2b_1$

     (b) $b_3 = b_2 - b_1$, $b_4 = 2b_1 - b_2$

9. (a) $X = \begin{bmatrix} 0 \\ 0 \\ 0 \end{bmatrix}$ (b) $X = \begin{bmatrix} 4t \\ \frac{5t}{2} \\ t \end{bmatrix}$

Supplementary Exercises (page 61)

1. $x' = \frac{3}{5} x + \frac{4}{5} y$, $y' = -\frac{4}{5} x + \frac{3}{5} y$

2. $x' = x \cos\theta + y \sin\theta$, $y' = -x \sin\theta + y \cos\theta$

3. 3 pennies, 4 nickels, 6 dimes   4. $x = 4$, $y = 2$, $z = 3$

5. Infinitely many if $a = 2$ or $a = -\frac{3}{2}$; none otherwise.

6. (a) $a \neq 0$, $b \neq 2$ (b) $a \neq 0$, $b = 2$

     (c) $a = 0$, $b = 2$ (d) $a = 0$, $b \neq 2$

7. $K = \begin{bmatrix} 0 & 2 \\ 1 & 1 \end{bmatrix}$ 8. mpn multiplications and mp(n - 1) additions

10. $a = 1$, $b = -2$, $c = 3$ 11. $a = 1$, $b = -4$, $c = -5$

Exercise Set 2.1 (page 73)

1. (a) 2   (b) -7   2. 57

3. 5   4. 0

5. 59   6. $k^2 - 4k - 5$

7. 0   8. 425

9. 104   10. $-k^4 - k^3 + 10k^2 + 9k - 21$

11. 275   12. -536

13. 120   14. 0

15. A cofactor expansion along the first row:

$$8 \begin{vmatrix} 4 & -6 \\ 7 & 2 \end{vmatrix} - 2 \begin{vmatrix} -3 & -6 \\ 1 & 2 \end{vmatrix} - 1 \begin{vmatrix} -3 & 4 \\ 1 & 7 \end{vmatrix}$$

A cofactor expansion along the last row:

$$1 \begin{vmatrix} 2 & -1 \\ 4 & -6 \end{vmatrix} - 7 \begin{vmatrix} 8 & -1 \\ -3 & -6 \end{vmatrix} + 2 \begin{vmatrix} 8 & 2 \\ -3 & 4 \end{vmatrix}$$

Exercise Set 2.2 (page 81)

1. (a) 6   (b) -16   (c) 0   (d) 0

2. -21   3. -5   4. -7   5. 18

6. -21   7. 6   8. $\frac{1}{6}$ 9. -2

10. 5   11. 10   12. 5   13. 10

Exercise Set 2.3 (page 88)

1. (a) $\begin{bmatrix} 29 & 11 & -19 \\ -21 & 13 & 19 \\ 27 & 5 & 19 \end{bmatrix}$ (b) $\begin{bmatrix} 29 & -21 & 27 \\ 11 & 13 & 5 \\ -19 & 19 & 19 \end{bmatrix}$ (c) $\dfrac{1}{152} \begin{bmatrix} 29 & -21 & 27 \\ 11 & 13 & 5 \\ -19 & 19 & 19 \end{bmatrix}$

2. $A^{-1} = \begin{bmatrix} -1 & \dfrac{8}{3} & \dfrac{5}{3} \\ -1 & 2 & \dfrac{4}{3} \\ -\dfrac{2}{3} & 1 & \dfrac{2}{3} \end{bmatrix}$

3. $A^{-1} = \begin{bmatrix} \dfrac{2}{3} & 0 & -\dfrac{1}{3} \\ -\dfrac{2}{9} & \dfrac{1}{9} & \dfrac{1}{9} \\ -\dfrac{1}{3} & 0 & \dfrac{1}{3} \end{bmatrix}$

4. $A^{-1} = \begin{bmatrix} -4 & 3 & 0 & -1 \\ 2 & -1 & 0 & 0 \\ -7 & 0 & -1 & 8 \\ 6 & 0 & 1 & -7 \end{bmatrix}$

5. $x_1 = 1,\ x_2 = 2$

6. $x = \dfrac{3}{11},\ y = \dfrac{2}{11},\ z = -\dfrac{1}{11}$

7. $x = \dfrac{26}{21},\ y = \dfrac{25}{21},\ z = \dfrac{5}{7}$

8. $x_1 = -\dfrac{30}{11},\ x_2 = -\dfrac{38}{11},\ x_3 = -\dfrac{40}{11}$

9. $x_1 = 3,\ x_2 = 5,\ x_3 = -1,\ x_4 = 8$

10. Cramer's rule does not apply.   11. $z = 2$

Supplementary Exercises (page 90)

1. $x' = \dfrac{3}{5} x + \dfrac{4}{5} y,\ y' = -\dfrac{4}{5} x + \dfrac{3}{5} y$

2. $x' = x \cos \theta + y \sin \theta,\ y' = -x \sin \theta + y \cos \theta$

5. $\cos \beta = \dfrac{c^2 + a^2 - b^2}{2ac}$, $\cos \gamma = \dfrac{a^2 + b^2 - c^2}{2ab}$

10. (b) $\dfrac{19}{2}$

CHAPTER 3

Exercise Set 3.1 (page 102)

1. (a) $(-3, -4, -8, 4)$ (b) $(53, 34, 49, 20)$ (c) $(-1, 2, 7, -10)$
    (d) $(-99, -84, -150, 30)$ (e) $(-63, -28, -21, -69)$ (f) $(2, 6, 15, -14)$

2. $\left(-\dfrac{7}{6}, -1, -\dfrac{3}{2}, -\dfrac{1}{3}\right)$ 3. $c_1 = 1, c_2 = 1, c_3 = -1, c_4 = 1$

5. (a) 5 (b) $\sqrt{11}$ (c) $\sqrt{14}$ (d) $\sqrt{48}$

6. (a) $\sqrt{73}$ (b) $\sqrt{14} + 3\sqrt{7}$ (c) $4\sqrt{14}$ (d) $\sqrt{1801}$

    (e) $\left(\dfrac{2}{\sqrt{6}}, 0, \dfrac{1}{\sqrt{6}}, \dfrac{1}{\sqrt{6}}\right)$ (f) 1

Exercise Set 3.2 (page 108)

1. (a) $Q = \begin{bmatrix} 1 & 0 \\ 0 & 1 \end{bmatrix}$ (b) $Q = \begin{bmatrix} \sqrt{3}/2 & -1/2 \\ 1/2 & \sqrt{3}/2 \end{bmatrix}$

    (c) $Q = \begin{bmatrix} \sqrt{2}/2 & -\sqrt{2}/2 \\ \sqrt{2}/2 & \sqrt{2}/2 \end{bmatrix}$ (d) $Q = \begin{bmatrix} 1/2 & -\sqrt{3}/2 \\ \sqrt{3}/2 & 1/2 \end{bmatrix}$

    (e) $Q = \begin{bmatrix} 0 & -1 \\ 1 & 0 \end{bmatrix}$ (f) $Q = \begin{bmatrix} -1 & 0 \\ 0 & -1 \end{bmatrix}$

2. (a) $Q = \begin{bmatrix} 1/2 & -\sqrt{3}/2 \\ \sqrt{3}/2 & 1/2 \end{bmatrix}$ (b) $Q^{-1} = \begin{bmatrix} 1/2 & \sqrt{3}/2 \\ -\sqrt{3}/2 & 1/2 \end{bmatrix}$

(c) $x' = \frac{1}{2} + \frac{5\sqrt{3}}{2}$

(d) $x = -\frac{1 + 5\sqrt{3}}{2}$

$y' = -\frac{\sqrt{3}}{2} + \frac{5}{2}$

$y = \frac{5 - \sqrt{3}}{2}$

(e) $3x'^2 - y'^2 = 4$

3. (a) $Q = \begin{bmatrix} \sqrt{3}/2 & -1/2 \\ 1/2 & \sqrt{3}/2 \end{bmatrix}$

(b) $Q^{-1} = \begin{bmatrix} \sqrt{3}/2 & 1/2 \\ -1/2 & \sqrt{3}/2 \end{bmatrix}$

(c) $x' = 2 - \sqrt{3}$
$y' = 1 + 2\sqrt{3}$

(d) $x = 3\sqrt{3} + \frac{5}{2}$

$y = 3 - \frac{5}{2}\sqrt{3}$

(e) $3x'^2 - y'^2 = 10$

4. (a) $Q = \begin{bmatrix} \sqrt{2}/2 & -\sqrt{2}/2 \\ \sqrt{2}/2 & \sqrt{2}/2 \end{bmatrix}$

(b) $Q^{-1} = \begin{bmatrix} \sqrt{2}/2 & \sqrt{2}/2 \\ -\sqrt{2}/2 & \sqrt{2}/2 \end{bmatrix}$

(c) $x' = \frac{5}{\sqrt{2}}$

(d) $x = -\frac{6}{\sqrt{2}}$

$y' = -\frac{3}{\sqrt{2}}$

$y = -\frac{6}{\sqrt{2}}$

(e) $x'^2 + 3y'^2 = 2$

5. $2x^2 - 2xy + 2y^2 = 5$

6. (a) $\begin{bmatrix} \cos\theta & -\sin\theta & 0 \\ \sin & \cos & 0 \\ 0 & 0 & 1 \end{bmatrix}$

(b) $\begin{bmatrix} 1 & 0 & 0 \\ 0 & \cos\theta & -\sin\theta \\ 0 & \sin\theta & \cos\theta \end{bmatrix}$

(c) $\begin{bmatrix} \cos\theta & 0 & \sin\theta \\ 0 & 1 & 0 \\ -\sin\theta & 0 & \cos\theta \end{bmatrix}$

7. $\left(\frac{\sqrt{2}}{2}, -\frac{\sqrt{2}}{2}\right), \left(-\frac{\sqrt{2}}{2}, -\frac{\sqrt{2}}{2}\right), \left(-\frac{\sqrt{2}}{2}, \frac{\sqrt{2}}{2}\right)$

8. $x'\cos\theta - y'\sin\theta = 0$

Exercise Set 3.3 (page 120)

1. (a) n = 2, m = 1   (b) n = 1, m = 2

    (c) n = 3, m = 1   (d) n = 1, m = 4

    (e) n = 3, m = 2   (f) n = 2, m = 3

    (g) n = 3, m = 3   (h) n = 2, m = 2

2. (a) $T(0,0) = (0,0,0)$ (b) $T(-1,2) = (-2, 5, -3)$

    (c) $T(a,b) = (ab,\ a^2 + b^2,\ a - b)$

3. (a) $(-7, 1)$, $(-10,10)$, $(0, 0)$

    (b)

    (c)

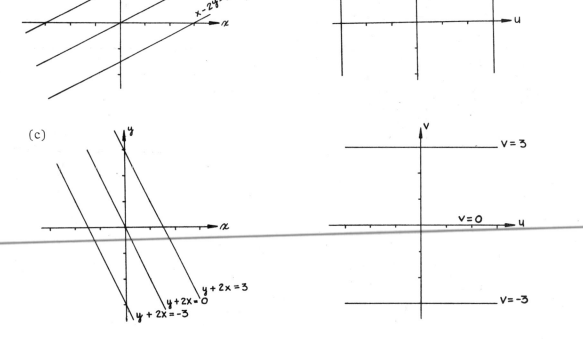

4. (a) $(8, -12)$, $(3, -8)$, $(0, -25)$

(b)

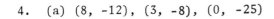

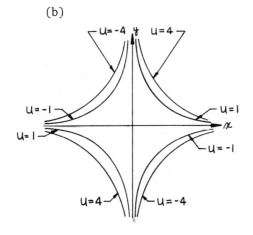

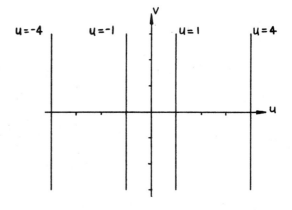

(c)

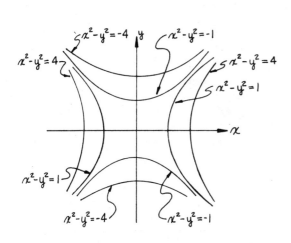

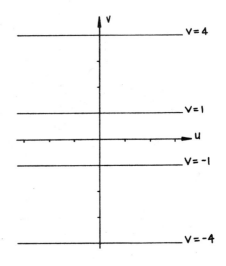

5. $xy \geq 0$

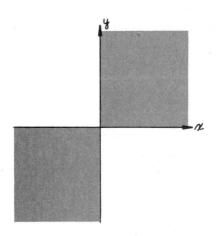

6. The entire x,y plane except (0,0).

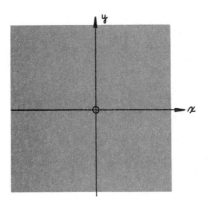

7. The entire plane except the lines
   $x + y = 1$, $x + y = -1$.

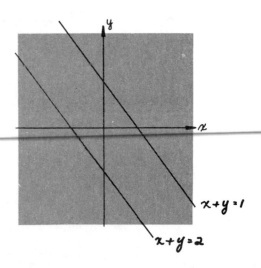

8. The entire xy plane.

10. $x = \dfrac{u + v}{4}$, $y = \dfrac{u - 3v}{4}$

11. Not 1-1.

12. Not 1-1.

13. $x = u^{1/3}$, $y = u^{1/3} - v$

14. Let: $x = au$, $y = bv$

15. (a) The rectangle in the uv-plane
     with vertices at: $(0,0)$, $(1,2)$
     $(3,-1)$, $(2,-3)$.

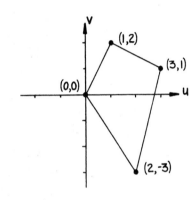

     (b) $16u + 6v = 35$

16. $3x - y = 5$

17. triangular region with vertices $(1,-2)$, $(2,-3)$, $(1,2)$.

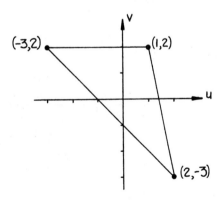

19. (a)

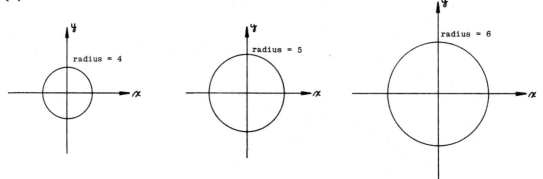

(b)

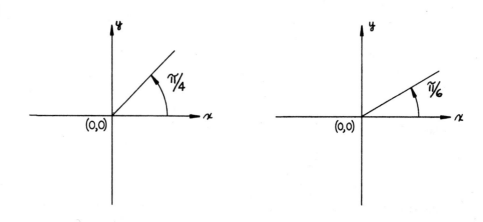

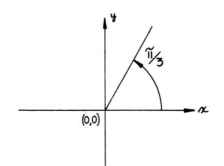

20. (a) right circular cylinder of radius 1 centered on the z-axis.

    (b) half plane making an angle of $\pi/4$ with the positive x-axis.

    (c) plane parallel to the xy-plane and 1 unit above it.

21. (a) A sphere, with radius r =1, centered at the origin.

    (b) A half plane making an angle of $\pi/4$ with the positive x-axis.

    (c) A cone centered on the positive z-axis and making an angle of $\pi/4$ with the positive z-axis.

NOTES

NOTES

NOTES

NOTES

NOTES

NOTES

NOTES

NOTES

NOTES

NOTES

NOTES